Y0-EIL-247

FLAMMABLE AND COMBUSTIBLE LIQUIDS CODE HANDBOOK

First Edition

FLAMMABLE AND COMBUSTIBLE LIQUIDS CODE HANDBOOK

FIRST EDITION

Based on the 1981 Edition of NFPA 30, the
Flammable and Combustible Liquids Code

Edited by

Martin F. Henry

Flammable Liquids Field Service Specialist

National Fire Protection Association
NFPA® Quincy, Massachusetts

Copyright © 1981
National Fire Protection Association
All Rights Reserved

NFPA No. SPP-58
ISBN: 0-87765-174-4
Library of Congress No.: 79-56909
Printed in U.S.A.

Dedication

This First Edition of the *Flammable and Combustible Liquids Code Handbook* is dedicated in memory of the late Edward C. Sommer who, for the past decade, served as Chairman of the Technical Committee responsible for the development of NFPA 30, *Flammable and Combustible Liquids Code*. He was also a past member of the NFPA's Board of Directors, and of its Flammable Liquids Advisory Committee.

Ed's contribution to the NFPA's consensus code-making process was unsurpassed. Beyond that, he was beloved by all those who knew him or who worked with him.

Edward C. Sommer

Contents

Foreword xi
Preface xiii
Acknowledgments xv

Chapter 1   General Provisions 1
  1-1   Scope and Application 1
  1-2   Definitions 5
  1-3   Storage 20
  1-4   Pressure Vessel 20
  1-5   Exits 21

Chapter 2   Tank Storage 23
  2-1   Design and Construction of Tanks 23
  2-2   Installation of Outside Aboveground Tanks 28
  2-3   Installation of Underground Tanks 74
  2-4   Installation of Tanks Inside of Buildings 82
  2-5   Supports, Foundations, and Anchorage for All Tank Locations 84
  2-6   Sources of Ignition 87
  2-7   Testing 88
  2-8   Fire Protection and Identification 89
  2-9   Prevention of Overfilling of Tanks 91
  2-10 Leakage Detection and Inventory Records for Underground Tanks .. 91

Chapter 3   Piping, Valves, and Fittings 95
  3-1   General 95
  3-2   Materials for Piping, Valves, and Fittings 96
  3-3   Pipe Joints 98
  3-4   Supports 99
  3-5   Protection Against Corrosion 99
  3-6   Valves 99
  3-7   Testing 100

Chapter 4   Container and Portable Tank Storage 101
  4-1   Scope 101
  4-2   Design, Construction, and Capacity of Containers 103
  4-3   Design, Construction, and Capacity of Storage Cabinets 105
  4-4   Design, Construction, and Operation of Separate Inside
      Storage Areas 107
  4-5   Indoor Storage 118
  4-6   Protection Requirements for Protected Storage of Liquids 128
  4-7   Fire Control 131
  4-8   Outdoor Storage 132

Chapter 5   Industrial Plants 139
   5-1   Scope 139
   5-2   Incidental Storage or Use of Liquids 140
   5-3   Unit Physical Operations 143
   5-4   Tank Vehicle and Tank Car Loading Unloading 146
   5-5   Fire Control 147
   5-6   Sources of Ignition 148
   5-7   Electrical Equipment 148
   5-8   Repairs to Equipment 150
   5-9   Housekeeping 150

Chapter 6   Bulk Plants 155
   6-1   Storage 155
   6-2   Buildings 156
   6-3   Loading and Unloading Facilities 157
   6-4   Wharves 161
   6-5   Electrical Equipment 165
   6-6   Sources of Ignition 165
   6-7   Drainage and Waste Disposal 166
   6-8   Fire Control 166

Chapter 7   Service Stations 173
   7-1   Scope 173
   7-2   Storage 173
   7-3   Piping, Valves, and Fittings 177
   7-4   Fuel Dispensing System 179
   7-5   Service Stations Located Inside Buildings 185
   7-6   Electrical Equipment 187
   7-7   Heating Equipment 187
   7-8   Operational Requirements 192
   7-9   Sources of Ignition 197
   7-10 Fire Control 197

Chapter 8   Processing Plants 199
   8-1   Scope 199
   8-2   Location 199
   8-3   Processing Building 200
   8-4   Liquid Handling 202
   8-5   Tank Vehicle and Tank Car Loading and Unloading 205
   8-6   Fire Control 205
   8-7   Sources of Ignition 207
   8-8   Housekeeping 211

Chapter 9   Refineries, Chemical Plants, and Distilleries 213
   9-1   Storage 213
   9-2   Wharves 213
   9-3   Location of Process Units 214
   9-4   Fire Control 214

Appendix A 217

Appendix B 225

Appendix C 229

Appendix D 237

Appendix E 243

Index 247

Foreword

For approximately 70 years the National Fire Protection Association has sponsored the *Flammable and Combustible Liquids Code.* Although originally known by other names, the *Code* is now proposed by the NFPA Committee on General Storage of Flammable Liquids, one of more than 170 technical committees operating within the framework of the NFPA standards-making activities.

The Technical Committee is made up of a group of well-qualified individuals representing a wide variety of interests. This group includes individuals involved with the manufacturing and sale of flammable and combustible liquids, with "users" or "consumers" of flammable and combustible liquids, and with those persons charged with the enforcement of the *Code.* The Committee fully recognizes the problems of developing a code that will be applicable both to large congested population areas and to small rural areas with only limited exposures to life or property. With these many variables in mind, it has been well accepted that this *Code* establishes reasonable requirements for the storage and handling of flammable and combustible liquids.

The NFPA recognizes that a code suitable for enforcement purposes must be concise and without explanatory text. Additionally, a code cannot be written to cover every situation that will be encountered. It must be applied with judgment and an awareness of the rationale for the requirements being enforced. A little help and counsel would make the application or use of the code easier and more understandable; therefore, the rationale for this *Flammable and Combustible Liquids Code Handbook.*

During 1978 and 1979, the NFPA delivered a number of three-day seminars on the *Code* in various locations across the country. The results proved mutually beneficial to the Association and to the users of NFPA 30. In keeping with the spirit and intent of consensus code making, the seminar students contributed heavily to the input for the 1981 Edition. They pointed out to us the sections of the *Code* that caused them difficulty in interpretation and understanding, and they recommended changes. The Technical Committee responsible for revisions to the *Code* responded by incorporating many revisions suggested to them by the seminar teaching staff. In addition, it became obvious that certain sections of the *Code* required explanatory background information, other sections needed to provide some rationale for certain *Code* provisions, and still others needed to offer additional guidance for users. This *Handbook* attempts to respond to those needs.

In keeping with the spirit of cooperative effort which has typified the development of both NFPA 30 and this *Handbook,* I invite you, the reader, to participate in future revisions. If you see shortcomings, ambiguities, or errors in the *Code,* submit proposals for change. If you want additional explanations in the *Handbook,* you have only to let the NFPA know.

The point was established at the seminars that NFPA 30 is your *Code.* You have the wherewithal to change it, but only if you participate in the process. I urge you to do so.

Miles E. Woodworth
Retired Secretary,
Flammable Liquids Code

Preface

As long ago as 1902, just six years into its existence, the National Fire Protection Association published a series of standards on the storage and use of kerosene and other flammable liquids. Several years later, action was undertaken to develop a single comprehensive ordinance dealing with requirements on the same subject. In 1913, that work was completed by the Committee on Laws and Ordinances and was adopted by the Association. It was published under the title *Suggested Ordinance Regulating the Use, Handling, Storage, and Sale of Inflammable Liquids and the Products Thereof.* And thus was NFPA 30 born.

Ten years later, it became apparent that developments in industry were such as to mandate a complete revision, particularly regarding large-scale flammable liquid storage. The 1926 Edition reflected that thinking. Membership on the Committee then (as now) included fire insurance companies, governmental authorities, oil industry, and other concerned interests. The intent of this group was the development of reasonable precautions to safeguard against the hazards connected with storage and use of flammable liquids. That intent still persists and prevails today.

The 1926 Standard was published in a form suitable for adoption as a city ordinance and for use as the basis of regulations. Enforcement might rest with the "Chief of the Fire Department," or any other enforcement agency, provided such agency was specifically identified.

The 1937 Edition saw the inclusion of the first informative appendix section, and it related to tanks in flooded regions. It was aptly entitled *Recommended Standards and Safe Practices for the Protection of Tanks Containing Flammable Liquids in Flooded Regions.* The 1981 Edition has updated and revised that section, and has incorporated it as a part of Chapter 2, Tank Storage. Incidentally, the 1937 Pamphlet Edition was available for just 15 cents.

Up until 1956, NFPA 30 was called a *Suggested Ordinance.* Beginning with the 1957 Edition it became known as the *Flammable Liquids Code,* and was recommended as the basis of legal regulation. In 1972, the *Code* was extended to include those liquids with flash points at or above 200°F (93.3°C). Until that time, it only applied to those liquids with flash points below 200°F (93.3°C).

The Committee which develops the *Code* has been at its task for nearly 70 years, contributing time, effort, and expertise. Of the approximately 30 members who now constitute the Committee, about one-half are members of the Society of Fire Protection Engineers. Three of the current members have an association with NFPA 30 which approaches a combined total of 100 years. All of the members bring a dedication to the same principles which governed the initiators of the project — namely, to develop guidelines for safety in the storage and handling of flammable liquids.

Edward C. Sommer
Chairman, Technical Committee on
General Storage of Flammable Liquids

Acknowledgments

Many persons contributed their time and knowledge to the preparation of this first edition of the FLAMMABLE AND COMBUSTIBLE LIQUIDS CODE HANDBOOK. The following persons, however, deserve special acknowledgment for their extensive work in the compilation, research, writing, and review of the materials contained herein.

William E. Doyle, P.E.
Oliver W. Johnson, Ph.D.
Ronald K. Melott, FPE (CA)

Cover design by K. Desmond.

Artwork for Figure 2-13 adapted with permission from Imperial Chemical Industries, Ltd.

Artwork within the commentary on the various sections of the *Flammable and Combustible Liquids Code* by K. Desmond and N. Maria.

NOTE: The text, illustrations, and captions to photographs that make up the commentary on the various sections of the *Flammable and Combustible Liquids Code Handbook* are printed in color. The photographs, all of which are part of the commentary, are printed in black for better clarity. The text of the *Code* itself is printed in black.

1

General Provisions

The serious student is encouraged to study NFPA 321, *Standard on Basic Classification of Flammable and Combustible Liquids*[1], Chapter 4 of Section 4 of the NFPA *Fire Protection Handbook*[2], NFPA *Flash Point Index of Trade Name Liquids*[3], NFPA 325M, *Fire Hazard Properties of Flammable Liquids, Gases, and Volatile Solids*[4], and NFPA 49, *Hazardous Chemicals Data*[5]. Details on specific chemicals and chemical characteristics are presented in these documents.†

1-1 Scope and Application.

1-1.1 This code applies to all flammable and combustible liquids except those that are solid at 100 °F (37.8 °C) or above.

A complete understanding of the scope and limitations of the *Code* is desirable when using it as a guide, and essential where it is referenced in laws or ordinances.

The words "flammable and combustible" have essentially the meanings of common usage, subject to the limitations given in Section 1-2 of the *Code*, and also in the NFPA *Fire Protection Handbook*, Chapter 4, Section 4.

The term "liquid" excludes materials having a vapor pressure exceeding 40 psig at 100 °F (37.8 °C). An example of such liquids — liquefied petroleum gas — is covered in NFPA 58.[6] Other standards cover liquids such as liquefied natural gas (NFPA 59A[7]) and liquid hydrogen (NFPA 50B[8]).

Solids having a melting point below 100 °F (37.8 °C) are covered because such solids may be liquid at some ambient temperatures, are easily ignited, and may escape from containers and spread or flow to reach ignition sources.

Solids having a melting point of 100 °F (37.8 °C) or above are excluded from the provisions of the *Code*. Although tables of physical properties of liquids and volatile solids may list them as having flash points and they behave like Class III liquids when melted, they may be shipped as solids in metal or other containers, or even in paper bags. Therefore, rules designed for materials which are normally liquid cannot apply.

---

†Available from National Fire Protection Association, Batterymarch Park, Quincy, MA 02269.

Some materials do not have a sharp dividing line between the liquid and solid states (e.g., asphalt). A liquid is defined in Section 1-2 as any material more fluid than 300 penetration asphalt. See Section 1-2, "Definitions," for an explanation of this concept.

1-1.2 Requirements for the safe storage and use of the great variety of flammable and combustible liquids commonly available depend primarily on their fire characteristics, particularly the flash point, which is the basis for the several classifications of liquids as defined in Section 1-2. It should be noted that the classification of a liquid can be changed by contamination. For example, filling a Class II liquid into a tank which last contained a Class I liquid can alter its classification, as can exposing a Class II liquid to the vapors of a Class I liquid via an interconnecting vapor line (*see 2-2.6.4 and 2-3.5.6*). Care shall be exercised in such cases to apply the requirements appropriate to the actual classification.

Flash point was selected as the basis of classification because it is related to volatility. A flash point in the normal atmospheric temperature range or below means that ignition can occur at ambient temperatures. At atmospheric temperatures above the flash point, enough vapor will be present to spread some distance from the surface of the liquid. The expression "low flash — high hazard" applies. However, some liquids that will not burn may have flash points if a small amount of flammable or combustible liquid is added. See comments under 4-1.2(e). Within a closed container, a liquid held at a temperature slightly above the flash point will produce a mixture which can be ignited to burn with explosive violence. At still higher temperatures, a release of vapors can spread, with dilution, to become ignitible some distance away. If the vapor is confined within the container, it can become "too rich" to be ignited.

Liquids with higher flash points are a lesser hazard because of a decreased chance of ignition and a decreased potential for vapor spread.

These considerations are the basis for the various classifications of liquids given in Section 1-2.

The preceding section of the *Code*, 1-1.2, also points out that contamination of a liquid with one of higher risk classification can change its characteristics. For anything other than a minor contamination with a lower flash point liquid, the classification of the mixture will become that of the low-flash component.

Fuel oil contaminated with gasoline is a good example. Fuel oil is, by itself, a combustible liquid. When contaminated with gasoline, the mixture can become flammable. Another example is carbon tetrachloride added to gasoline, so that the mixture does not exhibit a flash point. However, carbon tetrachloride has a lower boiling point (greater propensity for evaporation) than does gasoline. Over a period of time a flash point will be developed, and eventually reach that of the higher boiling fractions of gasoline.

GENERAL PROVISIONS

1-1.3 The volatility of liquids is increased by heating. When Class II or Class III liquids are heated above their flash points, ventilation and electrical classification may be necessary in the immediate area.

> Increasing the ambient temperature in the space where combustible liquids are stored or heating a combustible liquid above its flash point may create a hazardous atmosphere which requires the elimination of ignition sources. This hazardous condition normally occurs only in the space where the temperature of the vapors remains above the flash point of the liquid, although condensation of a mixture in or above the flammable range may create a flammable mist. Fuel oil No. 6, for example, normally has a flash point above 150°F (62.2°C), but, when heated above its flash point, the volatility of the liquid is increased and it assumes some characteristics of lower flash point liquids. Therefore, in the area of the heated fuel oil vapors, precautions applicable to the handling of flammable liquids would be required.

1-1.4 Additional requirements may be necessary for the safe storage and use of liquids which have unusual burning characteristics, which are subject to self-ignition when exposed to the air, which are highly reactive with other substances, which are subject to explosive decomposition, or have other special properties which dictate safeguards over and above those specified for a normal liquid of similar flash point classification.

> **NFPA 49,** *Hazardous Chemicals Data*[9], gives the fire hazard properties of many chemicals, including a number of reactive or otherwise unstable flammable liquids. This publication should be studied when a material having these special properties is encountered. Special protective measures, such as increased spacing between tanks, are outlined in appropriate places in Chapter 2.

1-1.5 In particular installations the provisions of this code may be altered at the discretion of the authority having jurisdiction after consideration of the special features such as topographical conditions, barricades, walls, adequacy of building exits, nature of occupancies, proximity to buildings or adjoining property and character of construction of such buildings, capacity and construction of proposed tanks and character of liquids to be stored, nature of process, degree of private fire protection to be provided and the adequacy of facilities of the fire department to cope with flammable or combustible liquid fires.

> Other physical characteristics and chemical properties of a liquid such as its boiling point, vapor pressure, specific gravity, water solubility, and toxicity may have to be considered in concert with the topography, building, and exposure conditions. Conditions which may allow spillage or percolation into water supplies, or vapor travel which might prevent the escape of persons from the area in the event of an accident, must be considered by the authority that enforces NFPA 30. Greater protection, or even prohibition of an installation,

may be necessary in densely populated neighborhoods or in areas where the public fire suppression capabilities are not sufficient to provide reasonable protection to the public. Only a very few flammable liquids are violently water-reactive. Where such liquids are stored, it will be prudent to follow precautions given in NFPA 49.

1-1.6 Existing plants, equipment, buildings, structures and installations for the storage, handling, or use of flammable or combustible liquids which are not in strict compliance with the terms of this code may be continued in use at the discretion of the authority having jurisdiction provided they do not constitute a recognized hazard to life or adjoining property. The existence of a situation which might result in an explosion or sudden escalation of a fire, such as inadequate ventilation of confined spaces, lack of adequate emergency venting of a tank, failure to fireproof the supports of elevated tanks, or lack of drainage or dikes to control spills may constitute such a hazard.

> Codes are designed to protect life and property. If exposure were limited to property alone, noncomplying conditions existing prior to *Code* adoption may be allowed to continue via the "grandfather's clause."† This approach is reasonable since codes are often developed as a result of fire loss experiences. Therefore, requiring compliance with laws which did not exist at the time of construction would be unreasonable, unless subsequent experience clearly shows a high degree of peril to life or hazardous exposure to another's property. The enumerated conditions in 1-1.6 which should be considered "recognized hazards" are new to the 1981 edition of the *Code*. They are included to serve as a guide to the authority having jurisdiction. Where such recognized hazards exist, a "grandfather's clause" exclusion would not have application. The recognized hazard constitutes ground for applying enforcement retroactively. It should be noted that the requirement prohibiting unprotected steel supports for aboveground tanks dates back to 1913 when the *Code* was first drafted.

1-1.7 This code shall not apply to:

1-1.7.1 Transportation of flammable and combustible liquids. These requirements are contained in the U.S. Department of Transportation regulations or in NFPA 385, *Recommended Regulatory Standard for Tank Vehicles for Flammable and Combustible Liquids*.

> Transportation is intended to include movement of flammable and combustible liquids by air, rail, truck, ship, and pipeline. The referenced documents are normally applied only to interstate shipments, with intrastate shipment being regulated by state or local laws which may also include the adoption of the referenced documents as local law.

---

†A clause creating an exemption based on previously existing circumstances.

1-1.7.2 Storage, handling and use of fuel oil tanks and containers connected with oil burning equipment. These requirements are covered separately in NFPA 31, *Standard for the Installation of Oil Burning Equipment.*

1-1.7.3 Storage of flammable and combustible liquids on farms and isolated construction projects. These requirements are covered separately in NFPA 395, *Standard for the Storage of Flammable and Combustible Liquids on Farms and Isolated Construction Projects.*

1-1.7.4 Liquids without flash points that can be flammable under some conditions, such as certain halogenated hydrocarbons and mixtures containing halogenated hydrocarbons. (*See NFPA 321, Basic Classification of Flammable and Combustible Liquids.*)

Examples of liquids without flash points which may be flammable under certain conditions (such as heating in a closed vessel) are: methyl bromide, dichloromethane, trichloroethane, and trichloroethylene.

1-1.7.5 Mists, sprays or foams. (*Except flammable aerosols in containers, which are included in Chapter 4.*)

Details on some operations which produce mists, sprays, or foams are given in NFPA 33, *Standard on Spray Application Using Flammable and Combustible Materials.*[10]

1-1.8 Installations made in accordance with the applicable requirements of standards of the National Fire Protection Association: NFPA 32, *Drycleaning Plants*; NFPA 33, *Spray Application Using Flammable and Combustible Materials*; NFPA 34, *Dip Tanks Containing Flammable or Combustible Liquids*; NFPA 35, *Manufacture of Organic Coatings*; NFPA 36, *Solvent Extraction Plants*; NFPA 37, *Installation and Use of Stationary Combustion Engines and Gas Turbines*; NFPA 45, *Fire Protection for Laboratories Using Chemicals*; and NFPA 56C, *Laboratories in Health-Related Institutions*, shall be deemed to be in compliance with this code.

NFPA publishes many standards which apply to specific hazards or processes, and compliance with a more specifically oriented standard takes the place of the less specific requirements of NFPA 30.

1-2 Definitions.

Aerosol. A material which is dispensed from its container as a mist spray or foam by a propellant under pressure.

This definition applies to the contained product, not the container or the propellant. (See definition of "Flammable Aerosol.")

Apartment House. A building or that portion of a building containing more than two dwelling units.

The definition is intended to include triplexes, fourplexes, townhouses, and condominiums, provided there are more than two dwelling units under the roof of the building and each dwelling unit has independent cooking and bathroom facilities.

Approved. Means "acceptable to the authority having jurisdiction."

NOTE: The National Fire Protection Association does not approve, inspect or certify any installations, procedures, equipment, or material nor does it approve or evaluate testing laboratories. In determining the acceptability of installations or procedures, equipment or materials, the authority having jurisdiction may base acceptance on compliance with NFPA or other appropriate standards. In the absence of such standards, said authority may require evidence of proper installation, procedure or use. The authority having jurisdiction may also refer to the listings or labeling practices of an organization concerned with product evaluations which is in a position to determine compliance with appropriate standards for the current production of listed items.

Satisfactory performance is usually proven by fire tests, actual fire experience, or both.

Assembly Occupancy. The occupancy or use of a building or structure or any portion thereof by a gathering of persons for civic, political, travel, religious or recreational purposes.

Atmospheric Tank. A storage tank which has been designed to operate at pressures from atmospheric through 0.5 psig.

The expression 0.5 psig means one half pound per square inch above prevailing atmospheric pressure. This is because an ordinary pressure gage indicates the difference in pressure between a container to which it is connected and the pressure of the surrounding atmosphere. This is in contrast to the expression "psia" — pounds per square inch absolute — in which the reference point is zero pounds absolute pressure. Except for the reference to psia in the definition of flammable liquids, all pressures in the *Code* **are gage pressures, psig.**

Authority Having Jurisdiction. The "authority having jurisdiction" is the organization, office or individual responsible for "approving" equipment, an installation or a procedure.

NOTE: The phrase "authority having jurisdiction" is used in NFPA documents in a broad manner since jurisdictions and "approval" agencies vary as do their responsibilities. Where public safety is primary, the "authority having jurisdiction" may be a federal, state, local or other regional department or individual such as a fire chief, fire marshal, chief of a fire prevention bureau, labor department, health department, building official, electrical inspector, or others having statutory authority. For insurance purposes, an insurance inspection department, rating bureau, or other insurance company representative may be the "authority having jurisdiction." In many circumstances the property owner or his designated agent assumes the role of the "authority having jurisdiction"; at government installations, the commanding officer or departmental official may be the "authority having jurisdiction."

Barrel. A volume of 42 U.S. gal (158.9 L).

The 42 gallon (gal) designation for a barrel is a petroleum industry measurement standard. A common mistake is to consider the drum and the barrel as equivalents; however, the drum is normally a volume of 55 U.S. gal. The Imperial (British and Canadian) gal is 20 percent larger than the U.S. gal.

Basement. A story of a building or structure having ½ or more of its height below ground level and to which access for fire fighting purposes is unduly restricted.

In any particular case, an interpretation of this definition can best be left to the authority having jurisdiction, who should know to what extent access for fire fighting will be required. Openings for the injection of gaseous extinguishing agents, water spray, or high-expansion foam may, in some cases, be considered sufficient. (See Figure 1-1.)

Boiling Point. The boiling point of a liquid at a pressure of 14.7 psia (760 mm). Where an accurate boiling point is unavailable for the material in question, or for mixtures which do not have a constant boiling point, for purposes of this code the 10 percent point of a distillation performed in accordance with ASTM D-86-72*, *Standard Method of Test for Distillation of Petroleum Products*, may be used as the boiling point of the liquid.

The boiling point of the liquid is the temperature of the liquid at which its vapor pressure equals the atmospheric pressure. Above this temperature the pressure of the atmosphere can no longer hold the liquid in a liquid state and bubbles begin to be evolved. The lower the boiling point of a liquid the greater the vapor pressure, and consequently the greater the fire hazard potential. The boiling point is affected by the elevation above sea level since the pressure of the atmosphere decreases at elevations above sea level and the boiling point is lowered. Conversely, the boiling point rises with an increase in pressure. For purposes of comparison, water boils at 212°F (100°C) at sea level and at 208°F (97.8°C) at 2,200 ft (671 m) elevation. Some liquids are mixtures with both high and low boiling points, and do not have a fixed boiling point. Gasoline is an example of this; its distillation range is about 100°F to 400°F (37.8°C to 204.4°C). The temperature at which 10 percent is evaporated will be around 150°F (65.6°C).

Boil-Over. The expulsion of crude oil (or certain other liquids) from a burning tank. The light fractions of the crude oil burn off producing a heat wave in

---
*Available from American Society for Testing and Materials, 1916 Race St., Philadelphia. PA, 19103.

8   FLAMMABLE AND COMBUSTIBLE LIQUIDS CODE HANDBOOK

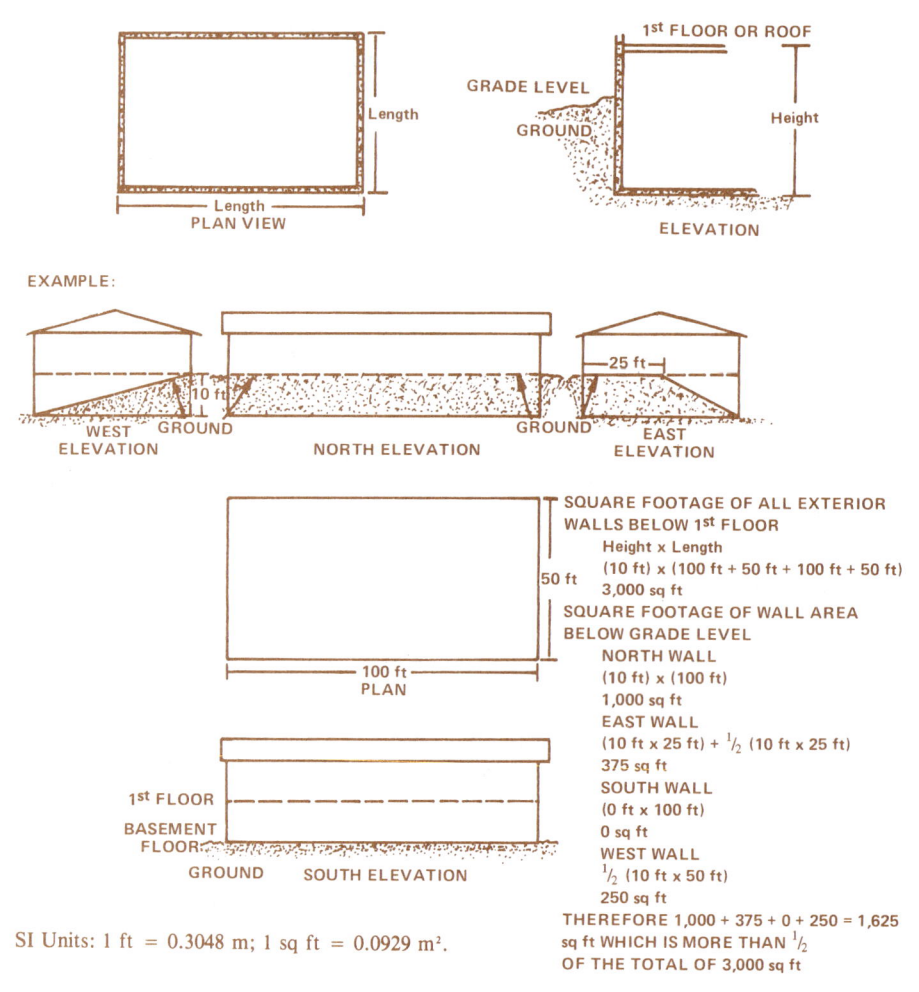

Figure 1-1. This example would not be considered a basement by this Code in spite of the square footage computation if there are adequate windows or doors on one or more sides such that fire fighting access could be made into all areas of this lower floor (basement) from the exterior. The definition of basement for this Code may differ from that of other NFPA codes and standards and adopted building codes. For purposes of this Code, access from the exterior for fire fighting purposes is the more critical factor.

the residue, which on reaching a water strata may result in the expulsion of a portion of the contents of the tank in the form of froth.

For a discussion of the mechanism of boil-over, see comments following 2-2.1.1.

Bulk Plant or Terminal.   That portion of a property where liquids are received by tank vessel, pipe lines, tank car, or tank vehicle, and are stored or blended in bulk for the purpose of distributing such liquids by tank vessel, pipe line, tank car, tank vehicle, portable tank, or container.

Chemical Plant. A large integrated plant or that portion of such a plant other than a refinery or distillery where liquids are produced by chemical reactions or used in chemical reactions.

Closed Container. A container as herein defined, so sealed by means of a lid or other device that neither liquid nor vapor will escape from it at ordinary temperatures.

This vessel shall not have any vents, either automatic, fixed, or pressure.

Combustible Liquids. See Liquids.

Container. Any vessel of 60 U.S. gal (227.1 L) or less capacity used for transporting or storing liquids.

This includes small cans, such as a 2-oz can of lighter fluid, and also the typical 55-gal drum.

Crude Petroleum. Hydrocarbon mixtures that have a flash point below 150°F (65.6°C) and which have not been processed in a refinery.

Technically, hydrocarbons are chemical compounds involving only the elements hydrogen and carbon; e.g., heptane $CH_3(CH_2)_5CH_3$; hexane $CH_3(CH_2)_4CH_3$. However, many of the materials covered in the *Code* also contain other elements such as oxygen, chlorine, nitrogen, etc.

Distillery. A plant or that portion of a plant where liquids produced by fermentation are concentrated, and where the concentrated products may also be mixed, stored, or packaged.

The process of distilling is used to remove impurities from a product or to separate products with different boiling points, and to concentrate desired components. It involves the vaporization of a liquid and condensation of the vapors. Fermentation is the process of chemically changing animal or vegetable matter by rearranging its elements or molecules through the use of bacteria, enzymes, yeasts, or molds.

Dwelling. A building occupied exclusively for residence purposes and having not more than two dwelling units or as a boarding or rooming house serving not more than 15 persons with meals or sleeping accommodations or both.

The 15 persons may or may not be related.

Dwelling Unit. One or more rooms arranged for the use of one or more individuals living together as a single house-keeping unit, with cooking, living, sanitary and sleeping facilities.

Educational Occupancy. The occupancy or use of a building or structure or any portion thereof by persons assembled for the purpose of learning or of receiving educational instruction.

Fire Area.† An area of a building separated from the remainder of the building by construction having a fire resistance of at least 1 hr and having all communicating openings properly protected by an assembly having a fire resistance rating of at least 1 hr.‡

> Such construction shall completely separate the area from all other building portions with fire-resistive construction from the floor through the roof (fire wall) or in such a manner as to *completely enclose* the area (top, bottom, and sides) if in a multistory building. The latter situation would also be true if construction is not provided to separate the attic, basement, crawl space, or lower floor area above or below the area to be separated. Thus, in a single-story building the fire wall must cut off the crawl space and attic area from the rest of the building, including the joist spaces. If not, then the ceiling and floor must be of rated construction, in addition to the walls. Fire walls cannot be penetrated by joists, beams, etc. (See Figure 1-2.)

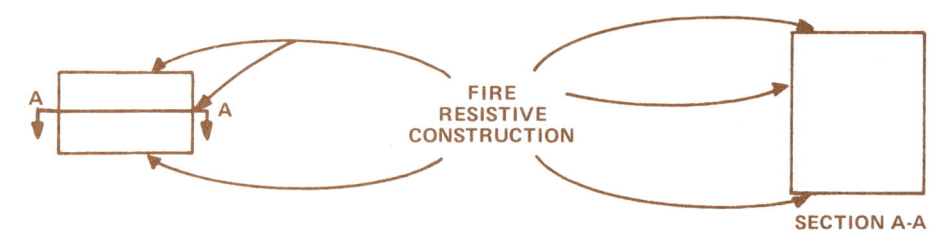

Figure 1-2. Fire-resistive construction.

Flammable Aerosol. An aerosol which is required to be labeled "Flammable" under the U.S. Federal Hazardous Substances Act.

See 4-1.3 for further treatment of flammable aerosols.

Flash Point. The minimum temperature at which a liquid gives off vapor in sufficient concentration to form an ignitible mixture with air near the surface of

---

†The reader is reminded that these definitions are applicable to this *Code* only and may differ from those in other codes and standards.

‡See NFPA 251, *Standard Methods of Fire Tests of Building Construction and Materials,*[11] or the building code for definitions of one-hour fire resistance.

the liquid within the vessel as specified by appropriate test procedure and apparatus as follows:

The flash point of a liquid having a viscosity less than 45 SUS at 100°F (37.8°C) and a flash point below 200°F (93.4°C) shall be determined in accordance with ASTM D-56-79,* *Standard Method of Test for Flash Point by the Tag Closed Tester.*

The flash point of a liquid having a viscosity of 45 SUS or more at 100°F (37.8°C) or a flash point of 200°F (93.4°C) or higher shall be determined in accordance with ASTM D-93-73,* *Standard Method of Test for Flash Point by the Pensky Martens Closed Tester.*

As an alternate, ASTM D-3243-73T, *Standard Methods of Tests for Flash Point of Aviation Turbine Fuels by Setaflash Closed Tester*, may be used for testing aviation turbine fuels within the scope of this procedure.

As an alternate, ASTM D-3278-73, *Standard Method of Tests for Flash Point of Liquids by Setaflash Closed Tester*, may be used for paints, enamels, lacquers, varnishes and related products and their components having flash points between 32°F (0°C) and 230°F (110°C), and having a viscosity lower than 150 stokes at 77°F (25°C).

As previously stated, the flash point is the primary characteristic for the classification of flammable and combustible liquids. Since it is the vapors of the flammable liquid which burn, vapor generation becomes a primary factor in determining the fire hazard. Flammable and combustible liquids having flash points below ambient storage temperatures generally have a rapid rate of flame spread, since it is not necessary for the heat of fire to expend its energy in heating the liquid to form additional vapors.[12]

Generally, the flash point of a substance is a few degrees below the fire point since at the flash point temperature the vapors are not being generated fast enough to sustain combustion. The flash point is normally several hundred degrees below the ignition temperature. (Fire point requires an outside ignition source. Ignition temperature requires no outside ignition source.) Thus, as the term *flash point* suggests, the vapors generated at that temperature will flash, but burning (fire point) will not continue.

Measurement of flash points has been made with a variety of methods. Originally, it was determined by heating the liquid in a small container open to the ambient air, and the flash point was the lowest temperature at which a small flame was observed to pass across the liquid surface when a test flame was applied to it. This is the "open-cup flash test." Subsequently, when more emphasis was placed on the ignition of vapors within a closed container (such as in tank explosions), the belief developed that closed-cup testing more accurately reflected the hazard of handling these liquids. Closed-cup test

---

*Available from American Society for Testing and Materials, 1916 Race St., Philadelphia, PA 19103.

methods had been used, but were not widely accepted until 1970. Currently, the NFPA recognizes only the four listed closed-cup methods. The most widely used is the Tag Closed Tester, ASTM D-56-79.[13] The three other approved test methods, all also standardized by ASTM, are used for special purposes, such as high-viscosity fluids and fluids that form a film on the surface. Experience has shown that the Tag Closed Tester has encountered difficulties with these properties.

Details of construction of these testers and the prescribed methods of performing the tests are contained in publications of the ASTM.† (See Figure 1-3.)

Hotel.‡ Buildings or groups of buildings under the same management in which there are sleeping accommodations for hire, primarily used by transients who are lodged with or without meals including but not limited to inns, clubs, motels and apartment hotels.

Institutional Occupancy.‡ The occupancy or use of a building or structure or any portion thereof by persons harbored or detained to receive medical, charitable or other care or treatment, or by persons involuntarily detained.

Labeled. Equipment or materials to which has been attached a label, symbol or other identifying mark of an organization acceptable to the "authority having jurisdiction" and concerned with product evaluation, that maintains periodic inspection of production of labeled equipment or materials and by whose labeling the manufacturer indicates compliance with appropriate standards or performance in a specified manner.

Liquid. For the purpose of this code, any material which has a fluidity greater than that of 300 penetration asphalt when tested in accordance with ASTM D-5-73*, *Test for Penetration for Bituminous Materials.* **When not otherwise identified, the term liquid shall mean both flammable and combustible liquids.**

As previously mentioned, the scope excludes liquids that are solid at 100°F (37.8°C) or above, thus creating a need to distinguish liquids and solids.

It was decided that any material that could spread or flow on a hot day should be considered a liquid. A material having the flow characteristics of 300 penetration asphalt was arbitrarily selected as being the most viscous material warranting classification as a liquid for the purpose of this *Code*; anything more viscous would be a solid, and hence not covered.

---

†Available from American Society for Testing and Materials, 1916 Race St., Philadelphia, PA 19103.

‡The reader is reminded that these definitions are applicable to this *Code* only and may differ from those in other codes and standards.

*Available from American Society for Testing and Materials, 1916 Race St., Philadelphia, PA 19103.

GENERAL PROVISIONS 13

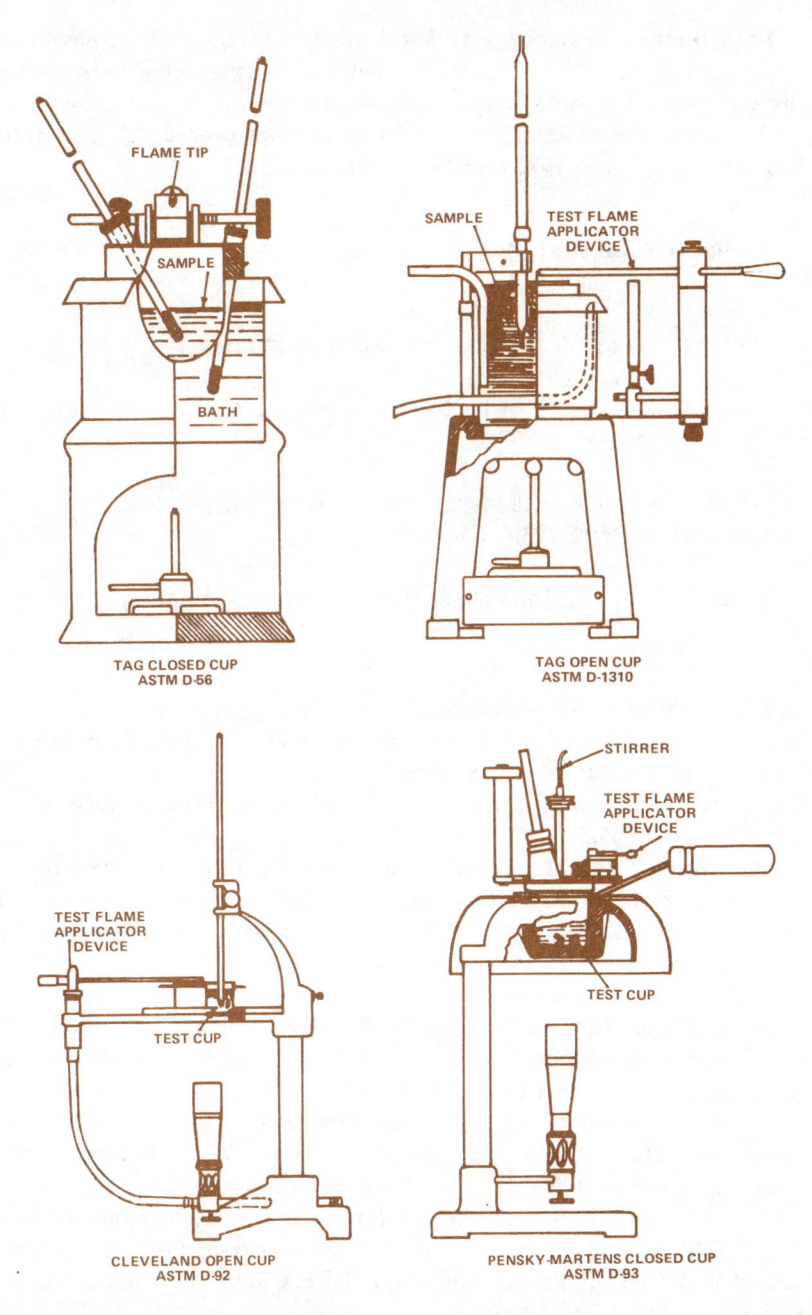

TAG CLOSED CUP
ASTM D-56

TAG OPEN CUP
ASTM D-1310

CLEVELAND OPEN CUP
ASTM D-92

PENSKY-MARTENS CLOSED CUP
ASTM D-93

Figure 1-3. Four commonly used testers for determining flash points of flammable or combustible liquids. The material to be tested is slowly heated, and at periodic intervals a test flame is applied to the vapor space. Flash point is the temperature at which a flash of fire is seen when the test flame is applied. More detailed descriptions for conducting tests for each type of testing apparatus are given in the applicable standard of the American Society for Testing and Materials or by the manufacturer.

Three hundred penetration asphalt is the most fluid grade of paving asphalt recognized by ASTM D-946-74.[14] Penetration is measured by recording the distance that a weighted, pointed rod penetrates in 5 sec into a sample at controlled temperature (see ASTM D-5-73 for detailed procedure). This definition first appeared in the 1972 revision of the *Code*.

Combustible Liquid. A liquid having a flash point at or above 100 °F (37.8 °C).

Combustible Liquids shall be subdivided as follows:

Class II liquids shall include those having flash points at or above 100 °F (37.8 °C) and below 140 °F (60 °C).

Class IIIA liquids shall include those having flash points at or above 140 °F (60 °C) and below 200 °F (93.4 °C).

Class IIIB liquids shall include those having flash points at or above 200 °F (93.4 °C).

The Department of Transportation (DOT) requires red labels on liquids having a flash point of 100 °F (37.8 °C) closed cup, or less. Many areas of the country reach ambient temperatures of 100 °F (37.8 °C). Under such circumstances, only moderate heating would be required to raise a liquid's temperature to 100 °F (37.8 °C).

The word "Combustible" first appeared in the title of the 1963 revision of the *Code*. The term Class III was applied to liquids having a flash point above 140 °F (60 °C) and below 200 °F (93.4 °C). Liquids having flash points above Class III were excluded from the scope.

In the 1966 revision, Class III was subdivided into Class IIIA and IIIB, the former being the same as the previous Class III, and the latter including liquids of all higher flash points. In spite of being defined, Class IIIB was excluded from the previous provision of the *Code*.

In the 1972 revision, the scope was broadened to include both Class IIIA and Class IIIB. The results of this change are minor, affecting principally some aspects of tank location and container storage.

Class II was assigned in 1963 to liquids having flash points at or above 100 °F (37.8 °C) and below 140 °F (60 °C). This corresponds to the solvent permitted in a Class II drycleaning plant (NFPA 32[15]). This definition has remained unchanged in both codes.

Flammable Liquid. A liquid having a flash point below 100 °F (37.8 °C) and having a vapor pressure not exceeding 40 pounds per square inch (absolute) at 100 °F (37.8 °C) shall be known as a Class I liquid.

Gases which are normally stored and shipped as liquids under pressure or refrigeration are not covered by this *Code*, since it is the stored pressure or refrigeration which causes the gases to liquefy, such as in the case of liquefied petroleum gases, liquefied hydrogen, and the like.

A vapor pressure of 40 psia (absolute) or 25.3 psig at 100 °F (37.8 °C) is the accepted dividing line between liquids and gases, except in the case of the Office of Pipeline Safety which calls liquefied petroleum gases, liquefied natural gas, and liquefied ammonia "highly volatile liquids."

Class I liquids shall be subdivided as follows:

Class IA shall include those having flash points below 73 °F (22.8 °C) and having a boiling point below 100 °F (37.8 °C).

Class IB shall include those having flash points below 73 °F (22.8 °C) and having a boiling point at or above 100 °F (37.8 °C).

Class IC shall include those having flash points at or above 73 °F (22.8 °C) and below 100 °F (37.8 °C).

Vapors from flammable and combustible liquids, in the pure state, are heavier than air. However, vapors in the pure state can exist only at or above the boiling point of the liquid. For all other conditions, the vapor is mixed with some air and the density is proportionately reduced. At the flash point temperature the vapor-air mixture will be less than about five percent vapor. The rest is air, and this mixture is only slightly heavier than uncontaminated air. Thus, dispersion of the mixture can easily result from wind or convection currents.

The 73 °F (22.8 °C) break on flash point temperature between Class IA and Classes IB and IC is based on the old Interstate Commerce Commission regulation which required a red label with the word "Flammable" on all liquids offered for interstate shipment which exhibited a flash point of below 80 °F (26.6 °C) open cup. (Note that the Consumer Product Safety Commission still uses this open-cup breaking point.) It was eventually agreed that 73 °F (22.8 °C) closed cup was the approximate equivalent of 80 °F (26.6 °C) open cup. The addition of the boiling point characteristic for distinguishing between Class IA and Class IB flammable liquids is included because the low boiling point liquids have a high vapor pressure which may approach the dividing line between liquids and gases [40 psia (25.3 psig)] and require storage in other than atmospheric tanks. NFPA 30 thus specifies greater requirements for those more hazardous materials. Under normal ambient temperatures, both Class IA and Class IB liquids generate sufficient vapors to create flammable vapor concentrations at all times. Pressure vessels or other strongly constructed containers capable of withstanding a designed internal pressure of as much as 5 psig must be used to store Class IA flammable liquids. Otherwise the product will evaporate from the container, creating a hazardous atmosphere as well as causing a loss of product.

The differentiation between Class IA and Class IB becomes important in selecting the type of tank, and in placing limits on quantities stored in buildings. See Chapter 4, Container and Portable Tank Storage.

Unstable (Reactive) Liquid. A liquid which in the pure state or as commercially produced or transported will vigorously polymerize, decompose, condense, or will become self-reactive under conditions of shock, pressure, or temperature.

Examples of unstable or reactive liquids are: acetaldehyde, acrylonitrile, butyl peracetate, ethyl acrylate, ethyl nitrite, hydroxylamine, methyl acrylate, nitromethane, peracetic acid, and styrene. See also discussion following 2-2.5.4.

Listed. Equipment or materials included in a list published by an organization acceptable to the "authority having jurisdiction" and concerned with product evaluation, that maintains periodic inspection of production of listed equipment or materials and whose listing states either that the equipment or material meets appropriate standards or has been tested and found suitable for use in a specified manner.

> NOTE: The means for identifying listed equipment may vary for each organization concerned with product evaluation, some of which do not recognize equipment as listed unless it is also labeled. The "authority having jurisdiction" should utilize the system employed by the listing organization to identify a listed product.

Low Pressure Tank. A storage tank designed to withstand an internal pressure above 0.5 psig (3.45 kPa) but not more than 15 psig (103.4 kPa).

Mercantile Occupancy. The occupancy or use of a building or structure or any portion thereof for the displaying, selling or buying of goods, wares, or merchandise.

Office Occupancy. The occupancy or use of a building or structure or any portion thereof for the transaction of business, or the rendering or receiving of professional services.

Portable Tank. Any closed vessel having a liquid capacity over 60 U.S. gallons (227.1 L) and not intended for fixed installation.

Pressure Vessel. Any fired or unfired vessel within the scope of the applicable section of the ASME Boiler and Pressure Vessel Code, available from American Society of Mechanical Engineers, United Engineering Center, 345 East 47th St., New York, NY 10017.

Protection for Exposures. Fire protection for structures on property adjacent to liquid storage. Fire protection for such structures shall be acceptable when located (1) within the jurisdiction of any public fire department, or (2) adjacent to plants having private fire brigades capable of providing cooling water streams on structures on property adjacent to liquid storage.

Exposure protection is dependent on a device or fire protection agency which can provide a barrier to the transmission of heat or a method for absorbing the heat generated by a fire. A public fire department will normally have sufficient capabilities to provide a hose stream(s) in the form of a water curtain, and the *Code* recognizes this capability. The larger plant fire brigades will normally have this capability also, but an extensive private water supply and hose system or a fire apparatus pumping engine will be required in addition to personnel.

Refinery. A plant in which flammable or combustible liquids are produced on a commercial scale from crude petroleum, natural gasoline, or other hydrocarbon sources.

Safety Can. An approved container, of not more than five gallons capacity, having a spring-closing lid and spout cover and so designed that it will safely relieve internal pressure when subjected to fire exposure.

The safety can is not designed or intended for use in areas where the periodic release of flammable vapors may create a hazardous atmosphere (such as in the trunk of an automobile). The main purposes of the safety can are the prevention of explosion of the container when it is heated, and the storing of liquids in a normally closed container. To accomplish this, a spring-operated cap on the pouring spout is provided, eliminating the need for a flame arrester in the spout. Even if the vapors coming past the spring-loaded cover should be in the flammable range, their velocity would be at least an order of magnitude greater than the intrinsic velocity of flame through the vapors, so a flashback would be impossible even if the vapors were ignited by an external source.

Separate Inside Storage Area. A room or building used for the storage of liquids in containers or portable tanks, separated from other types of occupancies. Such areas may include:

Inside Room. A room totally enclosed within a building and having no exterior walls.

Cutoff Room. A room within a building and having at least one exterior wall.

Attached Building. A building having only one common wall with a building having other type occupancies.

Exterior walls of cutoff rooms or attached buildings must provide some form of access for fire fighting purposes and, under certain circumstances, explosion venting features. (See 4-4.2.1.) Other details and construction requirements are set forth in Section 4-4. (See Figure 1-4.)

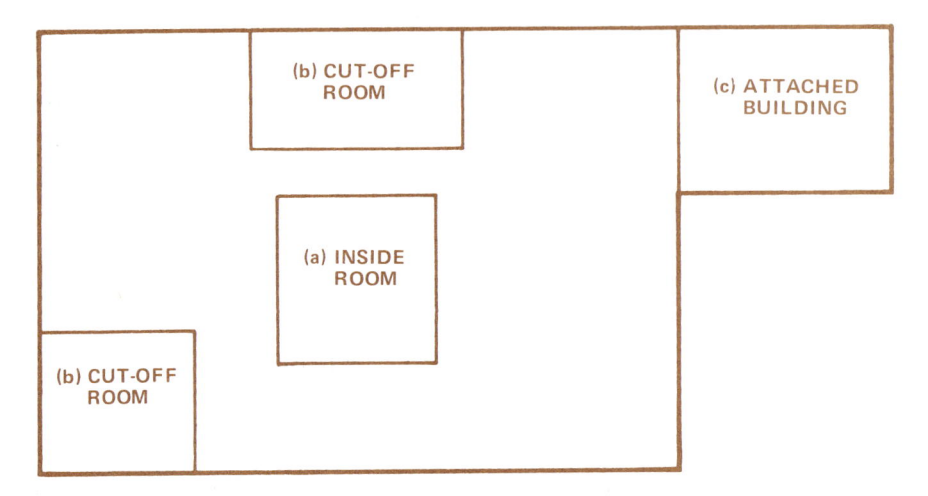

Figure 1-4. Diagram of: (a) inside room, (b) cutoff rooms, and (c) attached buildings.

Service Stations.

Automotive Service Station. That portion of property where liquids used as motor fuels are stored and dispensed from fixed equipment into the fuel tanks of motor vehicles and shall include any facilities available for the sale and service of tires, batteries and accessories, and for minor automotive maintenance work. Major automotive repairs, painting, body and fender work are excluded.

Marine Service Station. That portion of a property where liquids used as fuels are stored and dispensed from fixed equipment on shore, piers, wharves, or floating docks into the fuel tanks of self-propelled craft, and shall include all facilities used in connection therewith.

Service Station Located Inside Buildings. That portion of an automotive service station located within the perimeter of a building or building structure that also contains other occupancies. The service station may be enclosed or partially enclosed by the building walls, floors, ceilings, or partitions, or may be open to the outside. The service station dispensing area shall mean that area of the service station required for dispensing of fuels to motor vehicles. Dispensing of fuel at manufacturing, assembly, and testing operations is not included within this definition.

Vapor Pressure. The pressure, measured in pounds per square inch (absolute), exerted by a volatile liquid as determined by ASTM D323-72*, *Standard Method of Test for Vapor Pressure of Petroleum Products (Reid Method)*.

The vapor pressure is a measure of the liquid's propensity to evaporate or give off flammable vapors. The higher the vapor pressure the more volatile the liquid and, usually, the more readily the liquid gives off vapors. This vapor pressure is significant in determining whether a product is a liquid or a gas. Vapor pressure varies with temperature; it increases as the temperature of the liquid increases.

Vapor Processing Equipment. Those components of a vapor processing system which are designed to process vapors or liquids captured during filling operations at service stations, bulk plants, or terminals.

Vapor Processing System. A system designed to capture and process vapors displaced during filling operations at service stations, bulk plants, or terminals by use of mechanical and/or chemical means. Examples are systems using blower-assist for capturing vapors, and refrigeration, absorption and combustion systems for processing vapors.

The processing of vapors normally would include combustion of the vapors or the condensation of vapors back into the liquid state.

Vapor Recovery System. A system designed to capture and retain, without processing, vapors displaced during filling operations at service stations, bulk plants, or terminals. Examples are balanced-pressure vapor displacement systems and vacuum assist systems without vapor processing.

The basic function of a vapor recovery system is to prevent vapors from being released into the atmosphere either through returning the vapors to the container from which the liquid is being drawn or otherwise confining the vapors to prevent their release into the atmosphere.

Ventilation. As specified in this code, ventilation is for the prevention of fire and explosion. It is considered adequate if it is sufficient to prevent accumulation of significant quantities of vapor-air mixtures in concentration over one-fourth of the lower flammable limit.

Ventilation is vital to the prevention of flammable liquid fires and explosions. Ventilation is used to confine and remove vapor to a safe location, and also to dilute the concentration of vapor. It can be accomplished through natural or forced air movement.

The lower flammable limit is the minimum concentration of vapor to air below which propagation of a flame will not occur in the presence of an igni-

---

*Available from American Society for Testing and Materials, 1916 Race St., Philadelphia, PA 19103.

tion source. The upper flammable limit is the maximum vapor-to-air concentration above which propagation of flame will not occur. If a vapor-to-air mixture is below the lower flammable limit, it is "too lean"; if it is above the upper flammable limit, it is "too rich" to transmit flame.

When the *Code* describes ventilation as being adequate when vapor-air mixtures are not over one-fourth of the lower flammable limit, a safety factor of four-to-one is established.

When the vapor-to-air ratio is between the lower and upper flammable limits, ignition can occur and explosions may result. In the intermediate range between the upper and lower limits, the ignition is more intense and an explosion more violent.

Warehouses.

General Purpose Warehouse. A separate, detached building or portion of a building used only for warehousing-type operations.

NOTE: Warehousing operations referred to above are those operations not accessible to the public and include general purpose, merchandise, distribution and industrial warehouse-type operations.

Liquid Warehouse. A separate, detached building or attached building used for warehousing-type operations for liquids.

1-3 Storage. Liquids shall be stored in tanks or in containers in accordance with Chapter 2 or Chapter 4.

1-4 Pressure Vessel. All new pressure vessels containing liquids shall comply with 1-4.1, 1-4.2 or 1-4.3, as applicable.

1-4.1 Fired pressure vessels shall be designed and constructed in accordance with Section I (Power Boilers) 1971, or Section VIII, Division 1 or Division 2 (Pressure Vessels) 1974, as applicable, of the ASME *Boiler and Pressure Vessel Code**.

A fired pressure vessel is a container to which heat is applied through direct flame contact. Fired pressure vessels are mostly used to cause their contents to undergo an endothermic (heat absorbing) chemical reaction such as cracking of heavy oil to get lighter products such as gasoline. A pressure vessel is necessary because of the high vapor pressures created by applied heat.

1-4.2 Unfired pressure vessels shall be designed and constructed in accordance with Section VIII, Division 1 or Division 2, 1974 of the ASME *Boiler and Pressure Vessel Code**.

---

*Available from the American Society of Mechanical Engineers, United Engineering Center, 345 East 47th St., New York, NY 10017.

An unfired pressure vessel is one which, if heated, has the heat supplied by steam or other means not involving direct firing. Unfired pressure vessels are mostly used to contain liquefied gases, to contain reacting systems that may require some heat to start them, or, as in the case of nickel carbonyl, to contain highly toxic vapors.

Both fired and unfired pressure vessels may be used for flammable and combustible liquid storage, if desired.

1-4.3 Fired and unfired pressure vessels which do not conform to 1-4.1 or 1-4.2 may be used provided approval has been obtained from the state or other governmental jurisdiction in which they are to be used. Such pressure vessels are generally referred to as "State Special."

1-5 Exits.

1-5.1 Egress from buildings and areas covered by this code shall be in accordance with NFPA *101*®, the *Life Safety Code*®.

REFERENCES CITED BY *CODE*

(These publications comprise a part of the requirements to the extent called for by the **Code.***)*

NFPA 31, *Standard for Installation of Oil Burning Equipment*, NFPA, Boston, 1978.

NFPA 32, *Standard for Drycleaning Plants*, NFPA, Boston, 1979.

NFPA 321, *Basic Classification of Flammable and Combustible Liquids*, NFPA, Boston, 1976.

NFPA 33, *Standard for Spray Application Using Flammable and Combustible Materials*, NFPA, Boston, 1977.

NFPA 34, *Standard on Dipping and Coating Processes Using Flammable or Combustible Liquids*, NFPA, Boston, 1979.

NFPA 35, *Standard for the Manufacture of Organic Coatings*, NFPA, Boston, 1976.

NFPA 36, *Standard for Solvent Extraction Plants*, NFPA, Boston, 1978.

NFPA 37, *Standard for Stationary Combustion Engines and Gas Turbines*, NFPA, Boston, 1979.

NFPA 385, *Standard for Tank Vehicles for Flammable and Combustible Liquids*, NFPA, Boston, 1979.

NFPA 395, *Standard for the Storage of Flammable and Combustible Liquids on Farms and Isolated Construction Projects*, NFPA, Boston, 1980.

NFPA 45, *Standard on Fire Protection for Laboratories Using Chemicals*, NFPA, Boston, 1975.

NFPA 56C, *Standard for Laboratories in Health-Related Institutions*, NFPA, Boston, 1980.

NFPA 101, *Life Safety Code*, NFPA, Boston, 1981.

ASTM D-86-78, *Standard Method of Test for Distillation of Petroleum Products*, American Society for Testing and Materials, Philadelphia, 1978.

ASTM D-56-79, *Standard Method of Test for Flash Point by the Tag Closed Tester*, American Society for Testing and Materials, Philadelphia, 1979.

®Registered trademark of the National Fire Protection Association, Inc.

ASTM D-93-79, *Standard Method of Test for Flash Point by the Pensky Martens Closed Tester*, American Society for Testing and Materials, Philadelphia, 1979.

ASTM D-3243-77, *Standard Methods of Tests for Flash Point of Aviation Turbine Fuels by Setaflash Closed Tester*, American Society for Testing and Materials, Philadelphia, 1977.

ASTM D-3278-73, *Standard Method of Tests for Flash Point of Liquids by Setaflash Closed Tester*, American Society for Testing and Materials, Philadelphia, 1973.

ASTM D-5-73 (1978), *Test for Penetration for Bituminous Materials*, American Society for Testing and Materials, Philadelphia, 1978.

ASTM D-323-79, *Standard Method of Test for Vapor Pressure of Petroleum Products (Reid Method)*, American Society for Testing and Materials, Philadelphia, 1979.

ASME *Boiler and Pressure Vessel Code*, American Society of Mechanical Engineers, New York: Section I (Power Boilers) 1980; Section VIII, Division one and two, (Pressure Vessels) 1980.

REFERENCES CITED IN COMMENTARY

[1] NFPA 321, *Basic Classification of Flammable and Combustible Liquids*, NFPA, Boston, 1976.

[2] McKinnon, G.P., ed. *Fire Protection Handbook*, 14th ed., NFPA, Boston, 1976, Section 3, Chapter 3.

[3] *Flash Point Index of Trade Name Liquids*, NFPA, Boston, 1978.

[4] NFPA 325M, *Fire Hazard Properties of Flammable Liquids, Gases, and Volatile Solids*, NFPA, Boston, 1977.

[5] NFPA 49, *Hazardous Chemicals Data*, NFPA, Boston, 1975.

[6] NFPA 58, *Storage and Handling of Liquefied Petroleum Gases*, NFPA, Boston, 1979.

[7] NFPA 59A, *Liquefied Natural Gas, Production, Storage and Handling of*, NFPA, Boston, 1979.

[8] NFPA 50B, *Liquefied Hydrogen Systems at Consumer Sites*, NFPA, Boston, 1978.

[9] NFPA 49, *Hazardous Chemicals Data*, NFPA, Boston, 1975.

[10] NFPA 33, *Standard on Spray Application Using Flammable and Combustible Materials*, NFPA, Boston, 1977.

[11] NFPA 251, *Standard Methods of Fire Tests of Building Construction and Materials*, NFPA, Boston, 1979.

[12] ASTM E-502-74, *Standard Method of Test for Flash Point of Chemicals by Closed-Cup Methods*, American Society for Testing and Materials, Philadelphia, 1974.

[13] ASTM D-56-79, *Standard Method of Test for Flash Point by the Tag Closed Tester*, American Society for Testing and Materials, Philadelphia, 1979.

[14] ASTM D-946-74, *Specification for Asphalt Cement for Use in Pavement Construction*, American Society for Testing and Materials, Philadelphia, 1974.

[15] NFPA 32, *Standard for Drycleaning Plants*, NFPA, Boston, 1979.

2

Tank Storage

2-1 Design and Construction of Tanks.

2-1.1 Materials. Tanks shall be designed and built in accordance with recognized good engineering standards for the material of construction being used, and shall be of steel with the following limitations and exceptions:

> "Recognized good engineering standards" for the design and erection of tanks for the storage of flammable liquids have evolved from over 100 years of experience. No single standard is universally applicable because of the wide variety of liquids and their properties, and a latitude of choice of materials is permitted where circumstances warrant. The published standards listed in 2-1.3, 2-1.4, and 2-1.5 cover designs embodying a wide factor of safety to ensure that tanks will safely retain their contents under normal operating conditions. Emergency conditions, such as exposure to fire, require special details of construction or special protection. Such requirements are dealt with later.

(a) The material of tank construction shall be compatible with the liquid to be stored. In case of doubt about the properties of the liquid to be stored, the supplier, producer of the liquid, or other competent authority shall be consulted.

(b) Tanks constructed of combustible materials shall be subject to the approval of the authority having jurisdiction and limited to:

    1. Installation underground, or

    2. Use where required by the properties of the liquid stored, or

    3. Storage of Class IIIB liquids aboveground in areas not exposed to a spill or leak of Class I or Class II liquid, or

    4. Storage of Class IIIB liquids inside a building protected by an approved automatic fire extinguishing system.

(c) Unlined concrete tanks may be used for storing liquids having a gravity of 40 degrees API or heavier. Concrete tanks with special linings may be used for other services provided the design is in accordance with sound engineering practice.

> The expression "40 degrees API" is a measure of the specific gravity of the

liquid. Through long custom, the specific gravity of petroleum products has been designated in terms of gravity API, as measured with a hygrometer, calibrated so that specific gravity is related to degrees API by the formula:

$$\text{Degrees API} = \frac{141.5}{\text{sp.gr}} - 131.5$$

The API gravities for petroleum products may range from about 10° (for heavy lubricating oils) up to 80° or more (casing head gasoline). A low number (API) indicates a more dense, more viscous, and less volatile fluid. Normal motor grade gasoline has an API gravity in the range of 55° to 65°. The 40° gravity material mentioned in this paragraph corresponds roughly to kerosene or light stove distillate. Such materials have flash points in the range of 130°F (54.4°C) or above. Thus, a minor leakage from an unlined tank would not create an ignition hazard, and such tanks are permitted.

(d) Tanks may have combustible or noncombustible linings.

Tanks are occasionally lined with plastic material, such as sprayed-on epoxy resins, in the interest of corrosion protection or to offset small leaks. It doesn't matter whether the lining material is combustible or noncumbustible because the quantity is small and would not contribute to fire loss. Such a lining, even though electrically insulating, will not affect the risk of ignition from static electricity. (See NFPA 77.[1])

(e) Special engineering consideration shall be required if the specific gravity of the liquid to be stored exceeds that of water or if the tank is designed to contain liquids at a liquid temperature below 0°F (-17.8°C).

These considerations are embodied in API 650.[2]

2-1.2 Fabrication.

2-1.2.1 Tanks may be of any shape or type consistent with sound engineering design.

Sound engineering design can provide a structurally acceptable tank of any shape. However, as a practical matter, economy in fabrication, the material used, and the operating pressure limit the shape to one of the following:

1. Vertical cylindrical tanks with flat or nearly flat bottom and flat, coned, or domed roof.
2. Horizontal cylindrical tanks with flat or domed ends.
3. Spherical or toroidal tanks.

TANK STORAGE

In all cases the foundation and supporting structure must be considered as an integral part of the design. (See Section 2-5.)

2-1.2.2 Metal tanks shall be welded, riveted and caulked, or bolted, or constructed by use of a combination of these methods.

Tanks are constructed to be liquidtight up to the maximum filling level.

2-1.3 Atmospheric Tanks.

The term atmospheric tank is defined in Section 1-2 as a tank operating at pressures from atmospheric through 0.5 psig. *Welded Steel Tanks for Oil Storage*, (API 650),[2] uses the expression "internal pressures approximating atmospheric." For most large atmospheric tanks, the maximum permissible pressure will be somewhat less because it is considered undesirable that the roof plate welds be continuously stressed by pressure. A maximum pressure of about 1 in. (25.4 mm) of water or 0.6 oz/sq in. (0.25 kPa) is acceptable for large tanks.

2-1.3.1 Atmospheric tanks shall be built in accordance with recognized standards of design. Atmospheric tanks may be built and used within the scopes of the following:

(a) Underwriters Laboratories Inc., *Standard for Steel Aboveground Tanks for Flammable and Combustible Liquids*, UL142 — 1972; *Standard for Steel Underground Tanks for Flammable and Combustible Liquids*, UL58 — 1976; or *Standard for Steel Inside Tanks for Oil Burner Fuel*, UL80 — 1974.*

(b) American Petroleum Institute Standard No. 650, *Welded Steel Tanks for Oil Storage*, Sixth Edition, 1978.**

(c) American Petroleum Institute Specifications 12B, *Bolted Tanks for Storage of Production Liquids*, Twelfth Edition, January 1977**; 12D, *Field Welded Tanks for Storage of Production Liquids*, Eighth Edition, January 1977**; or 12F, *Shop Welded Tanks for Storage of Production Liquids*, Seventh Edition, January 1977**.

(d) Steel Tank Institute Standard No. STI-P3, *Specification for STI-P3 System of Corrosion Protection of Underground Steel Storage Tanks*, 1980***.

The exception for tanks storing crude oil in producing areas, 2-1.3.1(c), is in recognition of the fact that the tanks are usually small. An oil well may have a short useful life so that the need for tanks is temporary, and a bolted tank may

---

*Available from Underwriters Laboratories Inc., 333 Pfingsten Rd., Northbrook, IL 60062.

**Available from America Petroleum Institute, 2101 L St., N.W., Washington, DC 20037.

***Available from Steel Tank Institute, 666 Dundee Rd, Northbrook, IL 60062.

be easily dismantled and moved to a new location. Although such tanks are vulnerable to fire damage from a ground fire because of possible damage to the gaskets, the advantages outweigh the shortcomings. At main gathering stations and pipe line terminals, more substantial tanks would be required.

The Steel Tank Institute Standard is referenced to offer guidance on corrosion protection.

2-1.3.2 Low pressure tanks and pressure vessels may be used as atmospheric tanks.

This simply means that a tank designed for pressure will be acceptable for a less demanding service. However, the more stringent spacing requirements for pressure tanks specified in 2-2.1 are not to be waived unless the emergency venting system has the required capacity at atmospheric pressure.

2-1.3.3 Atmospheric tanks shall not be used for the storage of a liquid at a temperature at or above its boiling point.

In an atmospheric pressure tank, a liquid having a boiling point below prevailing atmospheric temperature would be expected to boil continuously, thus involving economic loss and increased risk of ignition from an external source. Boiling rate would actually depend upon the rate at which the tank receives heat from the atmosphere and from solar radiation. In interpreting this paragraph, it would be natural to consider boiling point to mean a boiling point at the atmospheric pressure prevailing at the tank location rather than at normal sea level pressure. For example, at an elevation of 10,000 ft (3,050 m) the contents of a tank might have a boiling point as much as 20 °F (11 °C) lower than at sea level. As a practical matter, this lowering of the boiling point is probably insignificant because the economic loss will usually dictate that liquids close to their boiling points be stored in pressure tanks.

2-1.4 Low Pressure Tanks.

2-1.4.1 The normal operating pressure of the tank shall not exceed the design pressure of the tank.

2-1.4.2 Low pressure tanks shall be built in accordance with recognized standards of design. Low pressure tanks may be built in accordance with:

(a) American Petroleum Institute Standard No. 620, *Recommended Rules for the Design and Construction of Large, Welded, Low-Pressure Storage Tanks*, Fifth Edition, 1973.**

(b) The principles of the *Code for Unfired Pressure Vessels*, Section VIII, Division I of the ASME *Boiler and Pressure Vessels Code*, 1974 Edition*.

---
*Available from the American Safety of Mechanical Engineers, United Engineering Center, 345 East 47th St., New York, NY 10017.
**Available from America Petroleum Institute, 2101 L St., N.W., Washington, DC 20037.

2-1.4.3 Tanks built according to Underwriters Laboratories Inc. requirements in 2-1.3.1 may be used for operating pressures not exceeding 1 psig (6.895 kPa) and shall be limited to 2.5 psig (17.24 kPa) under emergency venting conditions.

Tanks built to Underwriters Laboratories Inc. requirements (UL 142-1972)[3] are shop built and leak tested before shipment. Horizontal tanks are tested 5 to 7 psig, but are restricted in use to an operating pressure of 1 psig, and to 2.5 psig under emergency venting conditions. This restriction recognizes the manner of failure of a horizontal tank which is invariably accompanied by release of liquid. Vertical UL tanks are required to be tested to 1.5+ psig, which qualifies them as low pressure tanks. They are subject to liquid spill upon failure; thus, the 2.5 psig limitation under fire exposure conditions prevails. In contrast, a vertical tank built to UL 142 requirements and labeled "Built to Weak Shell-to-Roof Joint Design" would not be expected to fail in such a manner as not to release liquid. It is assumed that the weak seam will fail upon overpressure and that only vapors will be vented.

The term "emergency venting conditions" is not defined in Section 1-2. It is meant to describe a situation where a tank is fully exposed to flame, resulting in heating and boiling of the tank contents. Normal venting is based solely on liquid addition or withdrawal and atmospheric temperature and pressure changes. (See also 2-2.4 and 2-2.5.)

Since consequences of fire exposure can have a substantial effect on the permissible spacing and location of tanks as set forth in Section 2-2 and Tables 2-1 and 2-4, it is necessary to deal with this subject before commenting on specific requirements.

Flame contacting aboveground tanks can heat the liquid content, causing it to boil, and may also damage tank supports and the unwet portion of the tank shell. Boiling effects can be mitigated by design (2-2.5.3) or by additional relief valves (2-2.5.4 through 2-2.5.9). Supports for elevated tanks can be insulated to delay failure, or constructed of fire-resistive materials, as covered in Section 2-5.

Flame contact on the unwet portion of the shell of a steel tank can heat that portion of the shell of the tank not in contact with the liquid in the tank so that it loses much of its structural strength. For a vertical tank, this heating may result in distortion at the top of the shell, but collapse of the tank and spill of content is not likely. On the other hand, heating of the top of the steel shell of a horizontal tank is likely to result in structural failure, with release of contents.

For a pressure tank, the result of heating the unwet portion of the shell can be serious. Such tanks usually store liquids having boiling points below atmospheric temperature, and relief valves are set to maintain the resulting higher pressure. When heated sufficiently, the shell will lose strength and the resulting tear is likely to spread below the liquid level. This tear may extend completely around a horizontal tank so as to sever the head, release the contents, and cause the pieces to rocket. The possibility of this type of failure is recognized in Table 2-2.

2-1.4.4 Pressure vessels may be used as low pressure tanks.

2-1.5 Pressure Vessels.

2-1.5.1 The normal operating pressure of the vessel shall not exceed the design pressure of the vessel.

2-1.5.2 Storage tanks designed to withstand pressures above 15 psig shall meet the requirements of Section 1-4.

> The dividing line of 15 psig corresponds to the definition of a pressure vessel by ASME Code.[4] [See 2-1.4.2(b).]

2-1.6 Provisions for Internal Corrosion.

2-1.6.1 When tanks are not designed in accordance with the American Petroleum Institute, American Society of Mechanical Engineers or the Underwriters Laboratories Inc. Standards, or if corrosion is anticipated beyond that provided for in the design formulas used, additional metal thickness or suitable protective coatings or linings shall be provided to compensate for the corrosion loss expected during the design life of the tank.

> The choice of a suitable protective lining would depend on the nature of the liquid stored.

2-2 Installation of Outside Aboveground Tanks.

> The requirements of this section are intended to ensure that tanks are located so they will not jeopardize structures on the property of others. In the early days when tanks had wood-supported combustible roofs, it was not uncommon for fire to spread from one tank to another. Thus, regulations provided that tanks be widely spaced from each other and from all other facilities. With the advent of the steel roof, this risk greatly decreased. Experience indicated that, given no ground-spill fire, one tank could burn without damaging neighboring tanks or adjoining property. However, distance between tanks and spacing of tanks to property lines and adjacent structures remain principal planning criteria. Instead of restricting spacing on an arbitrarily selected distance, it is now considered more practical to base it on a fraction of tank diameter. Tables 2-1 through 2-4 employ this concept, while Tables 2-5 and 2-6 determine spacing by tank capacity.
>
> In Tables 2-1 through 2-4, spacing varies depending on the type of tank and roof, character of contents, fire protection or extinguishment facilities provided, and the protection for exposures. The term "protection for exposures" has a specialized meaning here, as defined in Section 1-2. It is recognized that it may be difficult or impossible to extinguish many tank fires. Therefore, efforts are made to protect any building or adjoining property from the fire. Fire protection for exposures should not be confused with installed equipment used

to extinguish the tank fire. Fire in tanks exceeding 150 ft (46 m) in diameter offers little possibility of extinguishment through fire fighting.

2-2.1 Location With Respect to Property Lines, Public Ways and Important Buildings on the Same Property.

2-2.1.1 Every aboveground tank for the storage of Class I, Class II or Class IIIA liquids, except as provided in 2-2.1.2 and those liquids with boil-over characteristics and unstable liquids, operating at pressures not in excess of 2.5 psig (17.24 kPa) and designed with a weak roof-to-shell seam or equipped with emergency venting devices which will not permit pressures to exceed 2.5 psig (17.24 kPa), shall be located in accordance with Table 2-1.

This paragraph introduces Table 2-1, states its scope, and by inference serves also as an introduction to Tables 2-2 through 2-7 which follow. In effect, it states that the spacings in Table 2-1 shall apply to all tanks storing Class I, Class II, and Class IIIA liquids, with the important exceptions discussed in 2-2.1.2 through 2-2.1.5 following.

This paragraph contains several terms having somewhat specialized meanings. The terms are used repeatedly in the text and in the tables in this chapter. These meanings must be understood and kept in mind to clarify the reasons for the provisions of this chapter. These terms are: "liquids with boil-over characteristics," "unstable liquids," "operating at pressures not in excess of 2.5 psig," "weak roof-to-shell seam," and "emergency venting devices." They are discussed in a slightly different order in the following paragraphs.

Boil-over. Tanks storing liquids having boil-over characteristics are not covered by Table 2-1, but are subjected to the more stringent requirements of Table 2-3. Any liquid will boil if heated sufficiently. Boiling can result in rise in liquid level because of the presence of the bubbles formed. Thus, a full vessel may overflow, releasing a small amount of liquid. Boiling ceases when the addition of heat ceases. In open-pool burning, the surface receives radiant heat to support the boiling. For most liquids a steady state condition results, and the burning proceeds without incident until all the oil is consumed. This is *not* a "boil-over."

The term boil-over is defined in Section 1-2, but an understanding of boil-over characteristics requires a more detailed discussion. To have "boil-over characteristics" an oil must have a wide range of boiling points, including a substantial proportion of volatile components, along with a highly viscous residue. This combination is present in most crude oils, but seldom in other petroleum products. A third requirement is the presence of at least a small amount of water-in-oil emulsion. When an open-top tank ignites, the following sequence of events occurs. As the fire burns, the surface layer is depleted of its more volatile components, becomes hotter and more dense than the original mixture, and sinks below the surface. It is replaced by a fresh layer of unevaporated oil. A gradually deepening layer of very hot oil results, having a temperature of 200°F (149°C) or more. When this hot layer reaches a previously settled layer of emulsion near the tank bottom, the water droplets in the emulsion are superheated to a temperature well above the boiling point at the prevailing pressure (atmospheric plus liquid head pressure). Boiling starts with almost explosive violence. The result is the expulsion of as much as half of the oil remaining in the tank, a sudden spread of burning oil over a wide area, and an overhead fireball which can produce serious injury to people in the area. This is a true boil-over, and constitutes the reason for the restrictions in tank locations given in 2-2.1.4 and Table 2-3. See also API 2021, *Guides for Fighting Fires In and Around Petroleum Storage Tanks*, Appendix B, Nov. 1974.[5]

Emergency Venting Devices. Emergency venting is covered in 2-2.5, but is nowhere defined in the *Code*. The word "emergency" in this sense means something unusual or beyond the conditions of normal operation, such as the exposure of a tank to surrounding fire. All of the tanks subject to the requirements of this *Code* contain a liquid which can burn. Fires seldom originate in a tank, and when they do the fire normally burns itself out without serious consequences. The emergency arises when liquid, spilled from a tank or its supporting pipe line, is ignited, producing what is commonly known as a "ground fire" or "spill fire." Contents of tanks exposed to such a fire are subject to boiling and evolution of vapor far in excess of that produced by normal operation. (See 2-2.4.) Emergency relief venting as described in 2-2.5 is required to accommodate this development. A weak roof-to-shell seam is one acceptable mechanism. This is also discussed following 2-1.4.2, in the comment concerning tanks built to UL requirements. The object of this type of construction is to ensure that a vertical tank will fail in such a way that no liquid will escape. It applies only to vertical tanks.

Unstable Liquids. Unstable liquids are defined in Section 1-2 (under "Liquid") and examples are given in the comment which follows the definition. Their unpredictable characteristics require special spacing (2-2.1.5 and Table 2-4).

Operating Pressures. The wording of 2-2.1.1 provides that Table 2-1 will apply to tanks with operating pressures up to 2.5 psig, provided they are equipped with emergency venting adequate to prevent pressures rising above this point. Tanks with operating pressures in excess of 2.5 psig are covered by Table 2-2. The reason for the substantially increased spacing requirements of Table 2-2 is that a tank at higher pressures is more likely to fail violently. A further restriction in the location of these tanks is given in 2-2.1.8.

For details of "weak roof-to-shell" construction, see UL 142 or API 620.[6] A tank which does *not* have this construction may have a roof-to-shell attachment involving a flanged shell or roof, or shell and roof riveted to an angle bar. See comment following 2-2.5.3.

(a) For the purpose of Section 2-2, a floating roof tank is defined as one which incorporates either:

1. A pontoon or double deck metal floating roof in an open top tank in accordance with API Standard 650, or

Experience has shown that tanks having floating roofs, as described in 2-2.1.1(a), are not likely to be involved in serious fire. Most fires in such tanks have burned only at the seal, and many have been easily extinguished. If the tank is of the open top type, hand-held extinguishing equipment may prove adequate. In cases where a floating roof has sunk, resulting in an open top tank fire, boil-over has not occurred. This is probably because the sunken roof presented a barrier to the heat wave. It is for these reasons that floating roof tanks are given preferred treatment in Table 2-3. For a further discussion of floating roofs, see Table 2-1.

2. A fixed metal roof with ventilation at the top and roof eaves in accordance with API Standard 650 and containing a metal floating roof or cover meeting any one of the following requirements:

a. A pontoon or double deck metal floating roof meeting the requirements of API Standard 650.

b. A metal floating cover supported by liquidtight metal pontoons or floats which provide sufficient buoyancy to prevent sinking of the cover when half of the pontoons or floats are punctured.

(b) An internal metal floating pan, roof or cover which does not meet the requirements of (a) 2., or one which uses plastic foam (except for seals) for flotation even if encapsulated in metal or fiber glass shall be considered as being a fixed roof tank.

Such construction is considered to be equivalent to a fixed roof tank because of the probability that the flotation devices would not withstand the heat of a fire, resulting in sinking of the roof.

2-2.1.2 Vertical tanks having a weak roof-to-shell seam and storing Class IIIA liquids may be located at one-half the distances specified in Table 2-1, provided the tanks are not within a diked area or drainage path for a tank storing a Class I or Class II liquid.

This relaxation of the spacing distance presented in Table 2-1 is in recognition of the fact that Class IIIA liquids cannot produce a flammable mixture in the vapor space at ordinary temperatures, coupled with the fact that a well-maintained tank with a weak roof-to-shell seam, in the absence of mechanical damage or an extremely violent earthquake, may not fail in such a way as to spill its contents.

2-2.1.3 Every aboveground tank for the storage of Class I, Class II or Class IIIA liquids, except those liquids with boil-over characteristics and unstable liquids, operating at pressures exceeding 2.5 psig (17.24 kPa), or equipped with emergency venting which will permit pressures to exceed 2.5 psig (17.24 kPa) shall be located in accordance with Table 2-2.

2-2.1.4 Every aboveground tank for storage of liquids with boil-over characteristics shall be located in accordance with Table 2-3.

(a) Liquids with boil-over characteristics shall not be stored in fixed roof tanks larger than 150 ft (46 m) diameter, unless an approved inerting system is provided on the tank.

> Liquids with boil-over characteristics are permitted in fixed roof tanks larger than 150 ft (46 m) in diameter if provided with an approved inerting system, thus decreasing the chance of ignition.
>
> The provision of an inerting system cannot prevent a boil-over, but it does decrease the chance that a fixed roof tank can become involved in fire.
>
> The term "approved inerting system" (also appearing in the "Protection" column of Table 2-4) describes a system whereby a tank is prevented from having within its vapor space a mixture which can be ignited during all phases of its operating cycle. Such a tank or vessel is termed "inerted."
>
> Inerting is defined in NFPA 69[7] to mean "the process of rendering a combustible mixture noncombustible by addition of an inert gas."
>
> The process of inerting an enclosed or nearly enclosed vessel, equipment, or room requires maintaining the oxygen concentration in the space low enough that combustion or other exothermic oxygenation reaction cannot be initiated under the most severe conditions of ignition. This requires that enough of the inert gas be added to maintain the oxygen content below about 11 or 12 percent (for limiting oxygen concentration for various flammable gases and vapors, see NFPA 69, Appendix B).
>
> The term "inert gas" as used does not mean chemically inert, but incapable of supporting combustion, and nonreactive with the contents of the system protected. Nitrogen, carbon dioxide, or a flue or stack gas containing less than 10 percent oxygen are commonly used. Helium and argon are effective, but too expensive except for highly specialized applications. Steam is not applicable unless it is practical to heat the whole system continuously above 160°F (71.1°C).
>
> An inerting system requires a continuously available source of inert gas as well as somewhat complicated controlled devices, and such an installation would not be undertaken in the absence of skilled supervision and urgent need. There is no established procedure for the "approval" of such a system, except to examine it carefully for conformance with all of the provisions of NFPA 69.

2-2.1.5 Every aboveground tank for the storage of unstable liquids shall be located in accordance with Table 2-4.

> An "unstable liquid," as defined in Section 1-2, is one that can undergo violent decomposition or semiexplosive reaction. This mandates a need for greater spacing than for tanks storing stable liquids. Unstable liquids are considered to present hazards akin to explosives. The tank would be unable to

contain the violent reaction caused by heating under fire conditions, and greater spacing from property which can be built upon is required than for any other class of liquid. (See Figure 2-1.)

Figure 2-1. Not all property can be built upon — for example, waterways, steep bluffs, railroad right-of-ways, and public streets and highways. This may have an effect on the location of tanks. It is possible that they could be located closer to property lines which cannot be built upon.

2-2.1.6 Every aboveground tank for the storage of Class IIIB liquids, excluding unstable liquids, shall be located in accordance with Table 2-5 except when located within a diked area or drainage path for a tank or tanks storing a Class I or Class II liquid. When a Class IIIB liquid storage tank is within the diked area or drainage path for a Class I or Class II liquid, 2-2.1.1 or 2-2.1.2 shall apply.

Class IIIB liquids are considered to be almost immune from accidental fire because of their high flash point, and distances from such tanks to the property line is minimal. However, if such tanks are located within the drainage area of Class I or Class II liquids, spill fires from other tanks could expose them.

Table 2-1 is based on the location restrictions given in the preceding paragraphs. It is more readily understood by first considering the headings of the four vertical columns, noting that as one progresses downward in each category in the column, the required spacing becomes progressively greater.

Type of Tank

Floating Roofs. The first tank category is for tanks having floating roofs. Experience has shown that open top tanks with floating roofs are not likely to

Table 2-1

Stable Liquids (Operating Pressure 2.5 psig or Less) (17.24 kPa)

Type of Tank	Protection	Minimum Distance in Feet from Property Line Which Is or Can Be Built Upon, Including the Opposite Side of a Public Way and Shall Be Not Less Than 5 Feet	Minimum Distance in Feet from Nearest Side of Any Public Way or from Nearest Important Building on the Same Property and Shall Be Not Less Than 5 Feet
Floating Roof [See 2-2.1.1(a)]	Protection for Exposures*	½ times diameter of tank	⅙ times diameter of tank
	None	Diameter of tank but need not exceed 175 feet	⅙ times diameter of tank
Vertical with Weak Roof to Shell Seam (See 2-2.1.1)	Approved foam or inerting system on tanks not exceeding 150 feet in diameter**	½ times diameter of tank	⅙ times diameter of tank
	Protection for Exposures*	Diameter of tank	⅓ times diameter of tank
	None	2 times diameter of tank but need not exceed 350 feet	⅓ times diameter of tank
Horizontal and Vertical with Emergency Relief Venting To Limit Pressures to 2.5 psig	Approved inerting system on the tank or approved foam system on vertical tanks	½ times Table 2-6	½ times Table 2-6
	Protection for Exposures*	Table 2-6	Table 2-6
	None	2 times Table 2-6	Table 2-6

SI Units: 1 ft = 0.3048 m.

*See definition for "Protection for Exposures."

**For tanks over 150 ft in diameter use "Protection for Exposures" or "None" as applicable.

be involved in fire because the only place where an ignitible mixture exists is in the narrow space above the oil surface within the seal itself.

Such tanks are unlikely to be ignited by flaming brands because such a brand would rarely fall close to the seal space. A floating roof tank is unlikely to be ignited by a direct stroke of lightning unless the roof is almost at the top of the tank. As first marketed, floating roof tanks were occasionally ignited by an electrical charge on the roof surface which was released coincident with a lightning stroke. This charge could escape from the roof to the shell, producing a spark in the seal space, and igniting a small fire around the rim of the roof. Corrective measures, currently followed to guard against such an incident, involve placing metallic conductors (shunts) between the rim of the floating roof and the metallic parts of the seal which bear against the tank shell. Such measures have seemed to be largely effective. In any event, a fire in the seal space of a floating roof tank can often be extinguished by hand extinguishers or by small equipment manually applied. Of still greater importance is the fact that if a floating roof sinks, it interferes with the thermal circulation of the oil and obstructs development of a boil-over.

For all of the preceding reasons, a vertical steel tank with a floating roof is given a preferred rating from the standpoint of exposing surrounding property to fire. This preferred rating applies if the floating roof is surmounted with a fixed roof with adequate ventilation.

Note that a floating roof containing combustible flotation components [2-2.1.1(b)] which are subject to destruction from a seal fire is no longer classified as a floating roof.

Fixed Roofs. A fixed roof tank is any tank that has a metallic roof other than a covered floating roof. Open top tanks are not recognized in the *Code.*

Other Tanks. The second category listed in column 1 of Table 2-1 covers vertical tanks with fixed roofs having weak roof-to-shell seams. These are assigned a slightly greater spacing requirement. Many large vertical tanks have an inherently weak roof attachment. Other tanks can have this feature incorporated in the design. (See Figure 2-2.)

The third category includes both horizontal and vertical tanks with emergency relief capabilities (other than a weak roof-to-shell attachment) capable of limiting the tank pressure to 2.5 psig under fire exposure conditions. In spite of adequate pressure relief, these tanks are not immune to failure if unwet portions of the shell are weakened by flame contact. Therefore, greater spacing requirements are assigned. Tanks which can be subjected to pressures greater than 2.5 psig under fire exposure conditions are covered by Table 2-2.

Tanks without some form of roof are not recognized in the *Code.*

Protection

Protection for exposures is defined in Section 1-2. The intent is that if fire should occur in the tank, there will be some fire fighting equipment available to prevent the ignition of adjacent property by cooling it with water. It is assumed that the fire can safely burn out and that no attempt will be made to extinguish the tank fire.

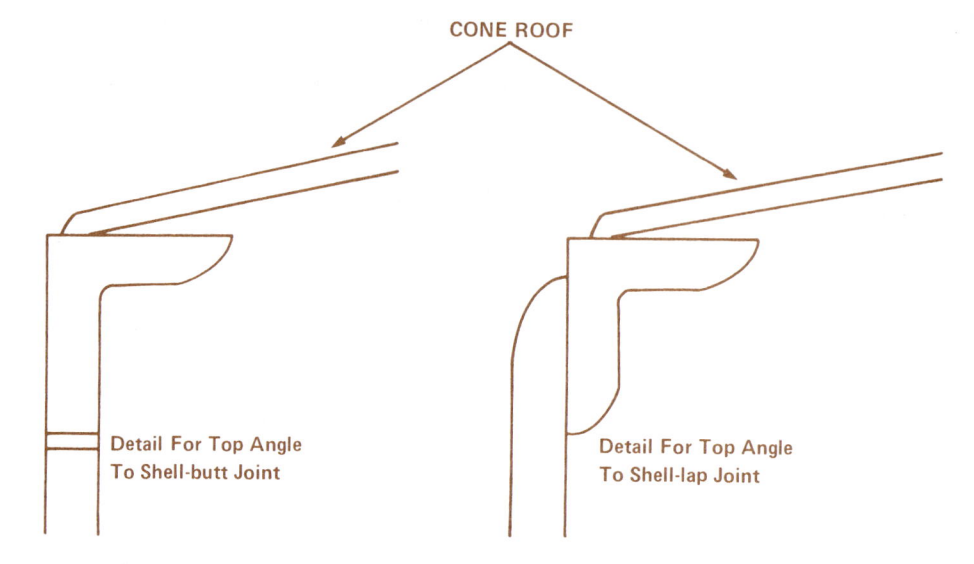

Figure 2-2. In a fire emergency, the design of a weak roof-to-shell seam allows the roof to tear free from the shell prior to the failure of any other seam. This type of construction is acceptable for relieving excessive internal pressure caused by exposure fires.

The next category concerns tanks equipped with an inerting system or with an approved foam extinguishing system — the latter being considered inapplicable for tanks exceeding 150 ft (46 m) in diameter because of logistic problems.

The third category is identical, except that the 150 ft (46 m) in diameter limitation has disappeared.

Minimum Distance to Property Lines

These two columns deal with slightly different situations. In the first, permanent structures can be built adjacent to the property line. In the second, intervening space is assured because of a public right of way, easement, railroad right of way, etc. In no case can a tank be placed less than 5 ft (1.5 m) from a property line. Spacing is prescribed as a function of tank diameter and tank construction, assuming that the effects of a fire will spread outside the limits of the owner's property, as influenced by the tank construction. The recognition of the type of liquid stored is covered in Sections 2-2, 2-3, and 2-4.

All of the preceding spacings have been carefully developed through fire experience over the 70 years or more since the *Code* was first conceived. Spacing distances to adjoining property, and also between tanks (Table 2-7), have occasionally been decreased over the years as the mechanism of fire spread has become better understood through years of experiment and experience.

Increased spacing for all types of tanks and all conditions is required bcause a tank operating at a pressure greater than 2.5 psig can create a more sudden and violent incident upon release of vapor than one operating at a lower pressure. Such an incident might include the failure of a tank by sudden tear-

Table 2-2

Stable Liquids (Operating Pressure Greater Than 2.5 psig) (17.24 kPa)

Type of Tank	Protection	Minimum Distance in Feet from Property Line Which Is or Can Be Built Upon, Including the Opposite Side of a Public Way	Minimum Distance in Feet from Nearest Side of Any Public Way or from Nearest Important Building on the Same Property
Any Type	Protection for Exposures*	1½ times Table 2-6 but shall not be less than 25 feet	1½ times Table 2-6 but shall not be less than 25 feet
	None	3 times Table 2-6 but shall not be less than 50 feet	1½ times Table 2-6 but shall not be less than 25 feet

SI Units: 1 ft = 0.3048 m.

*See Definition for "Protection for Exposures."

Table 2-3

Boil-over Liquids

Type of Tank	Protection	Minimum Distance in Feet from Property Line Which Is or Can Be Built Upon, Including the Opposite Side of a Public Way and Shall Be Not Less Than 5 Feet	Minimum Distance in Feet from Nearest Side of Any Public Way or from Nearest Important Building on the Same Property and Shall Be Not Less Than 5 Feet
Floating Roof [See 2-2.1.1(a)]	Protection for Exposures*	½ times diameter of tank	⅙ times diameter of tank
	None	Diameter of tank	⅙ times diameter of tank
Fixed Roof [See 2-2.1.4(a)]	Approved foam or inerting system	Diameter of tank	⅙ times diameter of tank
	Protection for Exposures*	2 times diameter of tank	⅔ times diameter of tank
	None	4 times diameter of tank but need not exceed 350 feet	⅔ times diameter of tank

SI Units: 1 ft = 0.3048 m.

*See definition for "Protection for Exposures."

ing of the roof or shell, as a result of heating and softening of the metal from direct flame contact.

The phenomenon of a "boil-over" was discussed following 2-2.1.1.

Boiling results when heat is transmitted to the contents of a tank by the flame-exposed shell. The tank is usually intact and may not be involved in fire, other than vapor burning at the vent. If the tank has a weak roof-to-shell connection or is adequately vented, no liquid spill is likely. The fire can burn out without serious incident.

By contrast, a *boil-over* results from fire in an open top tank, not from surrounding fire on the ground. A tank having a fixed roof cannot burn unless the roof has been removed, as by explosion. If the roof has been removed and the tank contains an oil having boil-over characteristics, one or more boil-overs will likely occur. The increased spacing required by Table 2-3 reflects the need for safeguarding adjoining structures against this contingency.

Tanks containing unstable liquids (as defined in Section 1-2) are unpredictable as to performance during fire exposure. For this reason, greater spacings are required.

Tanks storing Class IIIB liquids are rarely involved in fire, and spacing requirements are minimized.

2-2.1.7 Where two tank properties of diverse ownership have a common boundary, the authority having jurisdiction may, with the written consent of the owners of the two properties, substitute the distances provided in 2-2.2.1 through 2-2.2.6 for the minimum distances set forth in 2-2.1.

This simply says that, where owners agree, each may accept the risk from any of the tanks on the other's property to the same extent that he accepts the risks from his own.

2-2.1.8 Where end failure of horizontal pressure tanks and vessels can expose property, the tank shall be placed with the longitudinal axis parallel to the nearest important exposure.

This requirement is based on the fact that a horizontal pressure tank exposed to fire is likely to travel (rocket) axially upon failure. Application of this rule involves the difficult decision as to what structure constitutes the nearest important risk. An office building on the same property or any occupied buildings on adjoining property would both be considered "important exposures." A tank containing flammable liquids might also be so considered.

2-2.2 Spacing (Shell-to-Shell) Between Any Two Adjacent Aboveground Tanks.

2-2.2.1 Tanks storing Class I, II or IIIA stable liquids shall be separated in accordance with Table 2-7, except as provided in 2-2.2.2.

Table 2-4 Unstable Liquids

Type of Tank	Protection	Minimum Distance in Feet from Property Line Which Is or Can Be Built Upon, Including the Opposite Side of a Public Way	Minimum Distance in Feet from Nearest Side of Any Public Way or from Nearest Important Building on the Same Property
Horizontal and Vertical Tanks with Emergency Relief Venting to Permit Pressure Not in Excess of 2.5 psig	Tank protected with any one of the following: Approved water spray, Approved inerting, Approved insulation and refrigeration, Approved barricade	Table 2-6 but not less than 25 feet	Not less than 25 feet
	Protection for Exposures*	2½ times Table 2-6 but not less than 50 feet	Not less than 50 feet
	None	5 times Table 2-6 but not less than 100 feet	Not less than 100 feet
Horizontal and Vertical Tanks with Emergency Relief Venting to Permit Pressure Over 2.5 psig	Tank protected with any one of the following: Approved water spray, Approved inerting, Approved insulation and refrigeration, Approved barricade	2 times Table 2-6 but not less than 50 feet	Not less than 50 feet
	Protection for Exposures*	4 times Table 2-6 but not less than 100 feet	Not less than 100 feet
	None	8 times Table 2-6 but not less than 150 feet	Not less than 150 feet

SI Units: 1 ft = 0.3048 m.

*See definition for "Protection for Exposures."

Table 2-5 Class IIIB Liquids

Capacity Gallons	Minimum Distance in Feet from Property Line Which Is or Can Be Built Upon, Including the Opposite Side of a Public Way	Minimum Distance in Feet from Nearest Side of Any Public Way or from Nearest Important Building on the Same Property
12,000 or less	5	5
12,001 to 30,000	10	5
30,001 to 50,000	10	10
50,001 to 100,000	15	10
100,001 or more	15	15

SI Units: 1 ft = 0.3048 m; 1 gal = 3.785 L.

Table 2-6
Reference Table for Use in Tables 2-1 to 2-4

Capacity Tank Gallons	Minimum Distance in Feet from Property Line Which Is or Can Be Built Upon, Including the Opposite Side of a Public Way	Minimum Distance in Feet from Nearest Side of Any Public Way or from Nearest Important Building on the Same Property
275 or less	5	5
276 to 750	10	5
751 to 12,000	15	5
12,001 to 30,000	20	5
30,001 to 50,000	30	10
50,001 to 100,000	50	15
100,001 to 500,000	80	25
500,001 to 1,000,000	100	35
1,000,001 to 2,000,000	135	45
2,000,001 to 3,000,000	165	55
3,000,001 or more	175	60

SI Units: 1 ft = 0.3048 m; 1 gal = 3.785 L.

A minimum spacing between tanks storing stable liquids is given in Table 2-7. The minimum of 3 ft (0.91 m) is based on the need for access for maintenance and painting, and for application of cooling streams. Spacing for larger tanks is an arbitrary fraction of tank diameters, adequate to permit an orderly and safe arrangement for pipe lines, and to prevent spread of fire from one tank to another. Spacing alone is not a safeguard against fire spread from spilled liquid, control of which is covered in 2-2.3. Exceptions to the preceding

Table 2-7

Minimum Tank Spacing (Shell-to-Shell)

	Floating Roof Tanks	Fixed Roof Tanks Class I or II Liquids	Fixed Roof Tanks Class IIIA Liquids
All tanks not over 150 feet diameter	1/6 sum of adjacent tank diameters but not less than 3 feet	1/6 sum of adjacent tank diameters but not less than 3 feet	1/6 sum of adjacent tank diameters but not less than 3 feet
Tanks larger than 150 feet diameter			
If remote impounding is in accordance with 2-2.3.2	1/6 sum of adjacent tank diameters	1/4 sum of adjacent tank diameters	1/6 sum of adjacent tank diameters
If impounding is around tanks in accordance with 2-2.3.3	1/4 sum of adjacent tank diameters	1/3 sum of adjacent tank diameters	1/4 sum of adjacent tank diameters

SI Units: 1 ft = 0.3048 m.

follow. (For problem-solving situations concerning spacing requirements, see Figures 2-3 and 2-4.)

2-2.2.2 Crude petroleum tanks having individual capacities not exceeding 126,000 gal (3,000 barrels), when located at production facilities in isolated locations, need not be separated by more than 3 ft (0.91 m).

This exception is made because in a location without close neighbors it would be illogical to require large spacing between several tanks where a single tank storing the same quantity of oil would be permissible in view of lack of exposure to neighboring property. Thus, tanks may be placed close together, as a single risk, resulting in economics to piping and supervision, without increasing the risk to others.

2-2.2.3 Tanks used only for storing Class IIIB liquids may be spaced no less than 3 ft (0.91 m) apart unless within a diked area or drainage path for a tank storing a Class I or II liquid, in which case the provisions of Table 2-7 apply.

Close spacing for tanks storing Class IIIB liquids is permitted because of the low risk of ignition of the stored product.

2-2.2.4 For unstable liquids, the distance between such tanks shall not be less than one-half the sum of their diameters.

Spacing is required to be greater for the reason previously discussed under 2-2.1.5.

42 FLAMMABLE AND COMBUSTIBLE LIQUIDS CODE HANDBOOK

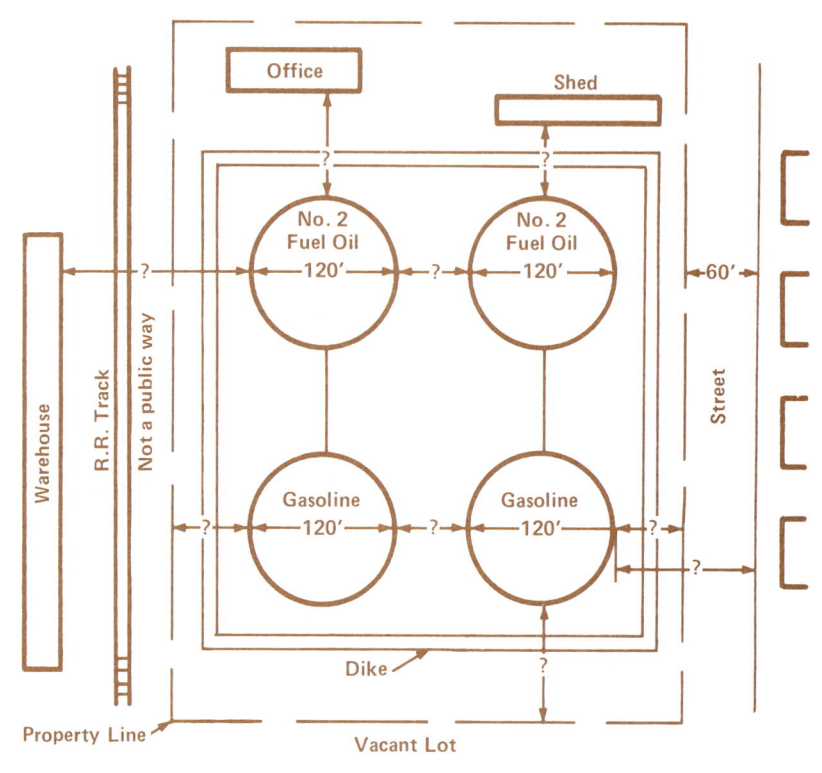

PROTECTION FOR EXPOSURES

SI Units: 1 ft = 0.3048 m.

Figure 2-3. DETERMINING REQUIRED DISTANCES (Problem 1). Assume that you have the responsibility for approving plans to add 4 additional tanks in a bulk plant. Two tanks will store No. 2 fuel oil, and 2 will store gasoline. Your problem is to ensure that the spacing requirements of NFPA 30 are followed. *Be advised that there is protection for exposures,* that all 4 tanks are vertical tanks built to API 650 specifications, are 120 ft (36.6 m) in diameter, and have weak roof-to-shell seams. You must determine the required distance from:
1. The gasoline tanks to the vacant lot.
2. The tanks to the near side of the street.
3. The tanks to the opposite side of the street. How does this affect the distance from the tank to the near side of the street in the preceding item 2? What options do you have to remedy the situation?
4. The fuel oil tank to the shed.
5. The fuel oil tank to the office building.
6. The tanks to the warehouse on the opposite side of the railroad tracks.
7. The tanks to the property line abutting the railroad right-of-way.
8. One tank to another; i.e., shell-to-shell spacing.

Solutions:

1. Table 2-1 indicates that, given a vertical tank with protection for ex-

posures and with weak roof-to-shell seams, the distance to the vacant lot (a property line which is or can be built upon) should be the diameter of the tank. The distance, then, should be 120 ft (36.6 m).

2. Table 2-1 indicates ⅓ the diameter of the tank from the near side of the street. In this case, that would mean 40 ft (12.19 m).

3. The distance to the opposite side of the street should be the diameter of the tank, or 120 ft (36.6 m). However, since the street is 60 ft (18.3 m) wide and the distance to the near side (in item 2) is just 40 ft (12.19 m), the plans must be altered in some way. There are a few options. If land availability is favorable, the tanks could be moved; the tanks could be decreased in diameter size, to 90 ft (27.4 m), for example; an approved foam system could be installed which would than lessen the requirement to ½ of the diameter, or 60 ft (18.3 m); finally, one might change the design of the tank to incorporate a floating roof, which would decrease the distance requirement to ½ of the tank diameter.

4. Given no additional pertinent information, one should assume that the shed constitutes an unimportant building on the company's premises. Therefore, there is no specified distance requirement. Naturally, the shed cannot be located within the diked area, so some minimum distance is implied, dependent on the plant configuration.

5. The office building must be considered to be an important structure on the same property, since it is an occupied building. Table 2-1 advises ⅓ of the tank diameter as a reasonable safe distance, therefore 40 ft (12.19 m) is the answer.

6. The warehouse constitutes a property line. Table 2-1 requires the separating distance to be the tank diameter, in this case 120 ft (36.6 m).

7. A railroad is not considered a public way. Therefore, Table 2-1 does not recommend any specific distance. In other words, there is no requirement for separation in the case of railroads.

8. Table 2-7 must be consulted to determine shell-to-shell spacing requirements. Since these tanks are less than 150 ft (45.72 m) in diameter, have fixed roofs, and are storing Class I and Class II liquids, the requirement calls for ⅙ of the sum of adjacent tank diameters. One-sixth of 240 (120 + 120) indicates a 40 ft (12.19 m) separation between adjacent tanks. Note that the adjacent tank distances are determined on a "one-to-one" basis. The word "adjacent" as used in the *Code* does not mean all 4 tanks in the diagram, but rather 1 tank adjacent to its nearest neighboring tank. The distances are computed in relation to 1 tank versus one neighboring tank, rather than in relation to all of the tanks taken collectively.

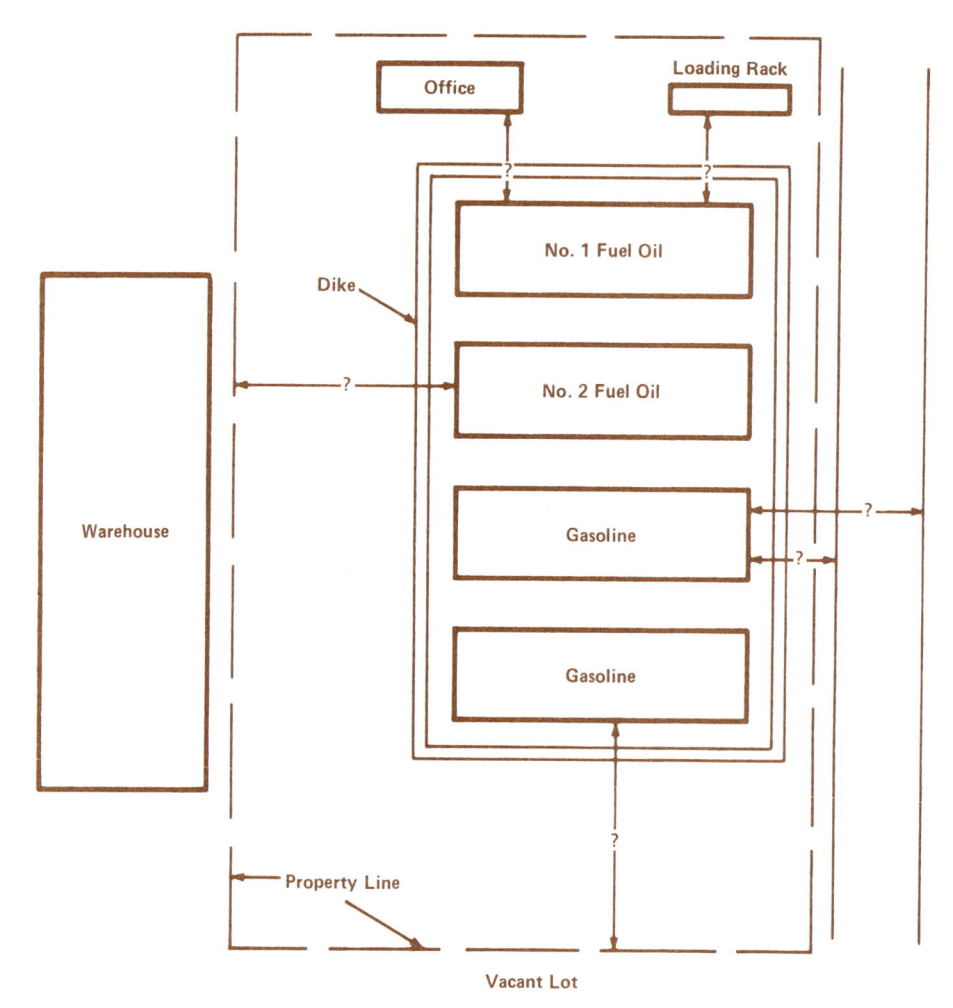

44   **FLAMMABLE AND COMBUSTIBLE LIQUIDS CODE HANDBOOK**

[Diagram showing a bulk plant layout with the following elements: Office and Loading Rack at top, tanks labeled "No. 1 Fuel Oil", "No. 2 Fuel Oil", "Gasoline", and "Gasoline" within a Dike, a Warehouse to the left, Property Line at bottom, and Vacant Lot below. Question marks ("?") indicate distances to be determined.]

ALL TANKS — 20,000 GALLONS CAPACITY
PROTECTION FOR EXPOSURES

SI Units: 1 gal = 3.785 L.

Figure 2-4. DETERMINING REQUIRED DISTANCES (Problem 2). Assume you are a County Fire Inspector responsible for the small bulk plant illustrated in this Figure. All 4 tanks are horizontal, equipped with emergency relief venting to limit pressures to 2.5 psig. There is protection for exposures. Each of the tanks has a capacity of 20,000 gal (75 700 L). In order to verify compliance with NFPA 30, you must determine the minimum distances required from:

1. *The gasoline tank to the property line adjoining the vacant lot.*
2. *The gasoline tanks to the near side and opposite side of the street.*
3. *The tank storing No. 1 fuel oil to the loading rack.*
4. *The tank storing No. 1 fuel oil to the office building.*
5. *The tank storing No. 2 fuel oil to the property line on which the warehouse is built.*

Solutions:

1. Table 2-1 refers the *Code* user to Table 2-6 for this situation. Table 2-6 indicates that the distance to the property line must be 20 ft (6.1 m).
2. Table 2-6 indicates 5 ft (1.5 m) from the nearest side of any public way, and 20 ft (6.1 m) from the opposite side of the public way.
3. The loading rack is certainly an important structure on the same property, and the distance indicated by Table 2-6 prescribes a minimum of 5 ft (1.5 m). However, another section of the *Code,* 6-3.1, gives specific distances for loading racks, namely 15 ft (4.6 m) for Class II and III liquids, and 25 ft (7.62 m) for Class I liquids. Since the loading rack very likely handles Class I and Class II liquids, it should be a distance of 25 ft (7.62 m) from aboveground tanks, warehouses, other plant buildings, or nearest line of adjoining property that can be built upon.
4. Table 2-6 indicates that a 5 ft (1.5 m) distance is required between the tank and the office building.
5. Table 2-6 requires tanks of the indicated capacity be spaced 20 ft (6.1 m) from the nearest property line.

NOTE 1. *If the preceding set of problems involved tanks on which the emergency venting permitted pressure to exceed 2.5 psig, Table 2-2 would be the reference to use. Table 2-2 refers the reader to Table 2-6, and indicates that the distances should be 1½ times those in Table 2-6, but not less than 25 ft (7.62 m).*

NOTE 2. *If the examples involved an unstable liquid with venting which limited pressures to 2.5 psig, Table 2-4 advises distances 2½ times those in Table 2-6, but not less than 50 ft (15.24 m).*

2-2.2.5 When tanks are in a diked area containing Class I or Class II liquids, or in the drainage path of Class I or Class II liquids, and are compacted in three or more rows or in an irregular pattern, greater spacing or other means may be required by the authority having jurisdiction to make tanks in the interior of the pattern accessible for fire fighting purposes.

2-2.2.6 The minimum horizontal separation between an LP-Gas container and a Class I, Class II or Class IIIA liquid storage tank shall be 20 ft (6.1 m) except in the case of Class I, Class II or Class IIIA liquid tanks operating at pressures exceeding 2.5 psig (17.2 kPa) or equipped with emergency venting which will permit pressures to exceed 2.5 psig (17.2 kPa) in which case the provisions of 2-2.2.1 and 2-2.2.2 shall apply. Suitable means shall be taken to prevent the accumulation of Class I, Class II or Class IIIA liquids under adjacent LP-Gas containers such as by dikes, diversion curbs or grading. When flammable or combustible liquid storage tanks are within a diked area, the LP-Gas containers shall be outside the diked area and at least 10 ft (3.05 m) away from the center line of the wall of the diked area. The foregoing provisions shall not apply when LP-Gas containers of 125 gal (473 L) or less capacity are installed adjacent to fuel oil supply tanks of 660 gal (2498 L) or less capacity. No horizontal separation is required between aboveground LP-Gas containers and underground flammable and combustible liquid tanks installed in accordance with Section 2-3.

2-2.3 Control of Spillage from Aboveground Tanks.

2-2.3.1 Facilities shall be provided so that any accidental discharge of any Class I, II or IIIA liquids will be prevented from endangering important facilities, adjoining property or reaching waterways, as provided for in 2-2.3.2 or 2-2.3.3. Tanks storing Class IIIB liquids do not require special drainage or diking provisions for fire protection purposes.

> The requirements of this paragraph are based on the rationale that release of product from a tank, however caused, shall not be permitted to endanger the important facilities and adjoining property, or to reach waterways where it could flow to navigable waters. The principal fear is that such a spill might be ignited and cause fire damage. Another major concern is environmental pollution which may be more extensive. Tanks storing Class IIIB liquids do not require spillage control for fire protection purposes alone because of the small chance that ignition could occur, although contamination of water will result whether or not a fire occurs. The word "storing" as it appears in the text does not limit application to standing storage, but also includes the acts of filling and emptying the tank.

2-2.3.2 Remote Impounding. Where protection of adjoining property or waterways is by means of drainage to a remote impounding area, so that impounded liquid will not be held against tanks, such systems shall comply with the following:

(a) A slope of not less than 1 percent away from the tank shall be provided for at least 50 ft (15.2 m) toward the impounding area.

(b) The impounding area shall have a capacity not less than that of the largest tank that can drain into it.

(c) The route of the drainage system shall be so located that, if the liquids in the drainage system are ignited, the fire will not seriously expose tanks or adjoining property.

(d) The confines of the impounding area shall be located so that when filled to capacity the liquid level will not be closer than 50 ft (15.2 m) from any property line that is or can be built upon, or from any tank.

> The concept here is that any spilled liquid will be conducted, by ditches or other channels, to an impounding area large enough to contain all of the liquid in the largest tank that can drain into it. The impounding area shall be so located that, if the spill ignited, the fire would not seriously damage other tanks or adjoining property. A reasonable slope away from the tank is required so that oil will not come to rest closer than 50 ft (15.2 m) from the tank or 50 ft (15.2 m) from any property line that can be built upon it. In early tank installations (where tanks were often placed on hills to achieve gravity flow), this objective was easily obtained. Locations on relatively flat terrain present a much more serious problem. It is, nonetheless, possible to achieve significant

drainage control with a minimum of grading. This is much preferred to the short-cut practice of placing each tank in the center of a flat diked area, where even a minor spill will make it likely that piping and control valves will be inaccessible and subject to fire damage. (See Figure 2-5.)

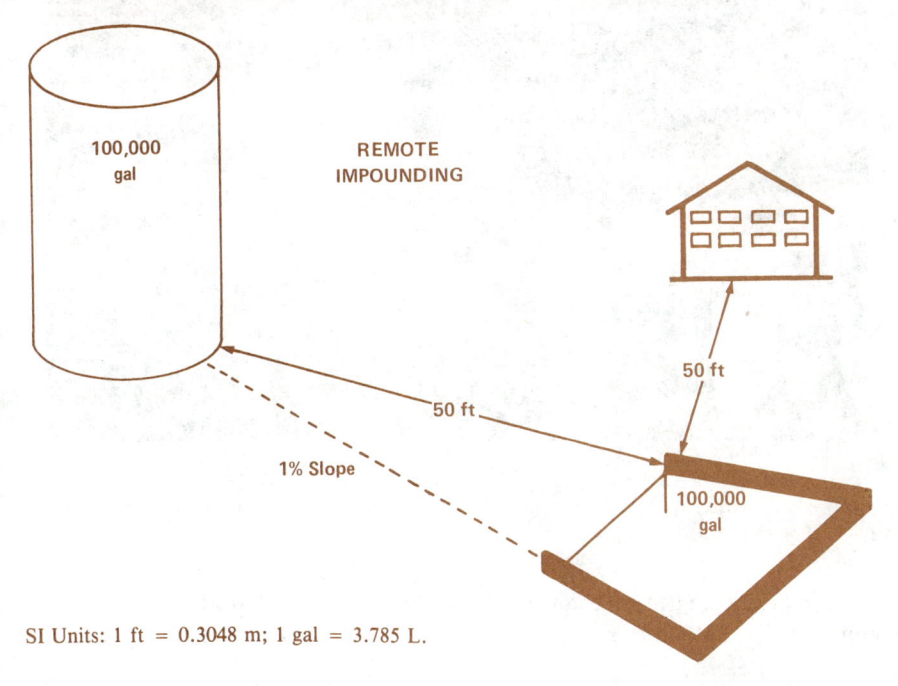

SI Units: 1 ft = 0.3048 m; 1 gal = 3.785 L.

Figure 2-5. The remote impounding area must be large enough to contain all of the liquid in the largest tank that can drain into it. A one percent slope away from the tank ensures adequate drainage control.

2-2.3.3 Impounding Around Tanks by Diking. When protection of adjoining property or waterways is by means of impounding by diking around the tanks, such system shall comply with the following:

Diking is less desirable than remote impounding because it may expose the leaking tank or adjacent tanks to a ground spill fire. (See Figure 2-6.)

(a) A slope of not less than 1 percent away from the tank shall be provided for at least 50 ft (15.2 m) or to the dike base, whichever is less.

A 1 percent slope for at least 50 ft (15.2 m) requires a 6 in. (152 mm) grade, since 1 ft per 100 ft (304 mm per 30 m) constitutes a slope of 1 percent.

(b) The volumetric capacity of the diked area shall not be less than the greatest amount of liquid that can be released from the largest tank within the diked area, assuming a full tank. To allow for volume occupied by tanks, the capacity of the

48   FLAMMABLE AND COMBUSTIBLE LIQUIDS CODE HANDBOOK

Figure 2-6.   There have been many fires which have involved all of the tanks in a single diked area when the fire has occurred on the ground. In these instances, the ground fire has caused all of the tanks to catch fire.

diked area enclosing more than one tank shall be calculated after deducting the volume of the tanks, other than the largest tank, below the height of the dike.

Application of this provision is best illustrated by problem solving:

What is the maximum volumetric dike capacity necessary for a 150 ft (46 m) diameter tank, 40 ft (12 m) high, within a diked enclosure?

1. Paragraph (b) states that the volumetric capacity of the dike enclosure must be capable of holding the greatest amount of liquid that can be released, assuming a full tank. The formula for the volume of a cylinder is

$$V = \frac{\pi d^2 \times h}{4}$$

where

V = volume of the tank in cu ft
π = 3.14
d = diameter of the tank
h = height of the tank

The solution gives the total volumetric capacity of the tank as **706,500 cu ft**. Therefore, the dike enclosure must be sufficiently large to contain that amount.

TANK STORAGE   49

The height of the diked wall will be dependent on the square footage available in the diked area. (See Figure 2-7, Problem 1.)

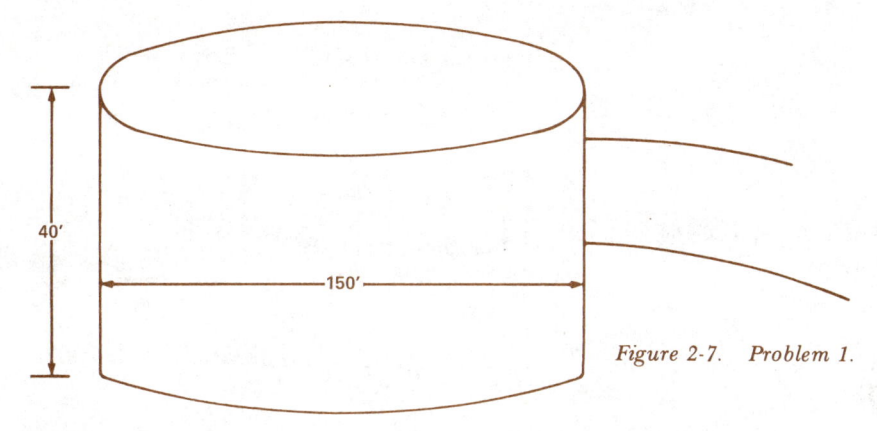

Figure 2-7. Problem 1.

SI Units: 1 ft = 0.3048 m.

2. Paragraph (b) also states that "the capacity of the diked area enclosing more than one tank shall be calculated after deducting the volume of the tanks, other than the largest tank, below the height of the dike." The following problem will exemplify the meaning:

> There are four tanks in a diked area. The largest is 100 ft (30 m) in diameter and 35 ft (10.6 m) high. Two other tanks are each 50 ft (15 m) in diameter and 20 ft (6 m) high, while the fourth tank is 40 ft (12 m) in diameter and 15 ft (4.5 m) high. The diked area is 70,000 sq ft (6503 m²) and has a 5 ft (1.5 m) high dike wall. Is the capacity of the dike sufficient for the largest tank?

Begin by determining the total volume of the diked area, which is equal to the area times the height of the dike wall; 70,000 × 5 = 350,000 cu ft (9912 m³) of volumetric capacity. (See Figure 2-8, Problem 2.)

We must now deduct from this volumetric capacity the volume taken up or consumed by the three smaller tanks. Keep in mind that they occupy available volumetric dike space only as high as the dike wall, or 5 ft (1.5 m). The value h in our volume formula, therefore, will be 5 ft (1.5 m).

The total volume of the smaller tanks is 25,905 cu ft (734 m³) [up to the 5 ft (1.5 m) level]. This volume, subtracted from the overall dike capacity, leaves us 324,095 cu ft (9178 m³) of true volumetric capacity.

We must now determine the volume of the largest tank. Our calculations tell us that the tank has a capacity of 274,750 cu ft (7781 m³). We know, then, that the diked volume is adequate.

(c) To permit access, the outside base of the dike at ground level shall be no closer than 10 ft (3.05 m) to any property line that is or can be built upon.

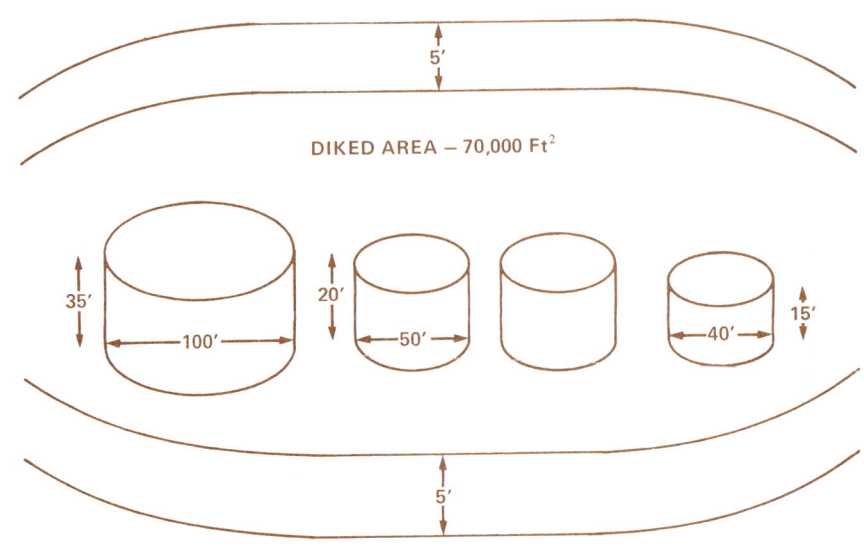

Figure 2-8. Problem 2.

SI Units: 1 ft = 0.3048 m.

This is a relatively new section of the *Code*, and is designed to permit access for fire fighting and to provide additional protection for buildings on adjoining property in the event of a fire in the dike area. (See Figure 2-9.)

Figure 2-9. Fire fighters operating within a diked area.

(d) Walls of the diked area shall be of earth, steel, concrete or solid masonry designed to be liquidtight and to withstand a full hydrostatic head. Earthen walls

3 ft (.91 m) or more in height shall have a flat section at the top not less than 2 ft (.61 m) wide. The slope of an earthen wall shall be consistent with the angle of repose of the material of which the wall is constructed. Diked areas for tanks containing Class I liquids located in extremely porous soils may require special treatment to prevent seepage of hazardous quantities of liquids to low-lying areas or waterways in case of spills.

Well-compacted clays will resist the permeation of liquid.

(e) Except as provided in (f) below, the walls of the diked area shall be restricted to an average interior height of 6 ft (1.8 m) above interior grade.

The reason for limiting the general dike height to 6 ft (1.8 m) is to provide an escape route for fire fighters engaged in combatting small fires within the dike. Exceptions to this requirement are allowed, but other precautions are then mandated. (See Figure 2-10.)

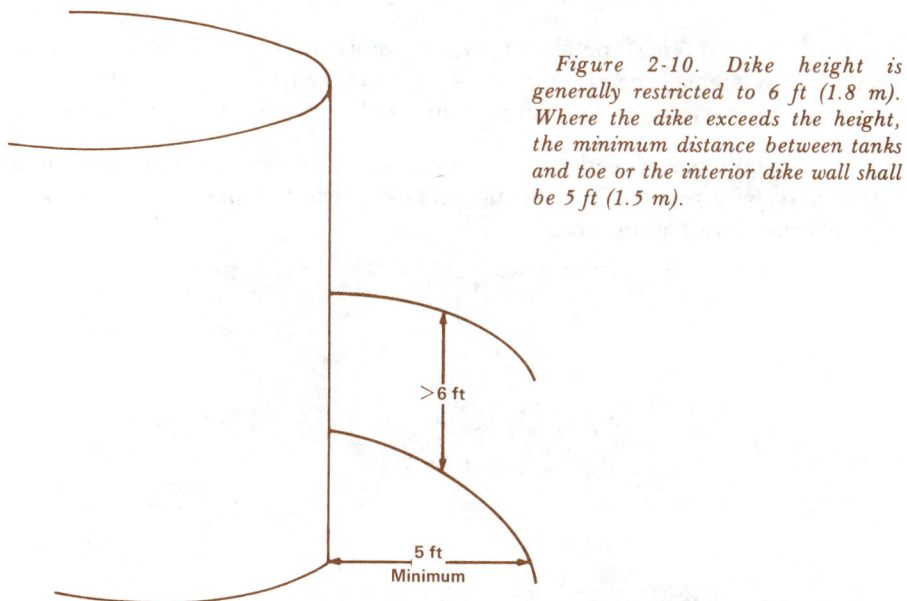

Figure 2-10. Dike height is generally restricted to 6 ft (1.8 m). Where the dike exceeds the height, the minimum distance between tanks and toe or the interior dike wall shall be 5 ft (1.5 m).

SI Units: 1 ft = 0.3048 m.

(f) Dikes may be higher than an average of 6 ft (1.8 m) above interior grade where provisions are made for normal access and necessary emergency access to tanks, valves and other equipment, and safe egress from the diked enclosure.

The reasons for dike heights in excess of 6 ft (1.8 m) involve land costs or land availability in built-up areas, and regulations in some jurisdictions requiring individual diking of tanks.

1. Where the average height of the dike containing Class I liquids is over 12 ft (3.6 m) high, measured from interior grade, or where the distance between any tank and the top inside edge of the dike wall is less than the height of the dike wall, provisions shall be made for normal operation of valves and for access to tank roof(s) without entering below the top of the dike. These provisions may be met through the use of remote operated valves, elevated walkways or similar arrangements.

2. Piping passing through dike walls shall be designed to prevent excessive stresses as a result of settlement or fire exposure.

3. The minimum distance between tanks and toe of the interior dike walls shall be 5 ft (1.5 m).

These provisions reflect a concern for Class I liquid vapors reaching unsafe concentrations when confined in the small space between the dike wall and the tank. Remotely operated valves or elevated walkways would eliminate the need for someone to enter the bottom of the diked area to operate a valve.

It should be emphasized that the provisions of (1), (2), and (3) apply only to the high dikes permitted in (f).

(g) Each diked area containing two or more tanks shall be subdivided, preferably by drainage channels or at least by intermediate curbs in order to prevent spills from endangering adjacent tanks within the diked area as follows:

This requirement is designed to control those relatively small spills which, in the past, have resulted in major fire and destruction of all tanks within a dike enclosure. (See Figure 2-11.)

Figure 2-11. These intermediate dikes are required to hold at least 10 percent of the capacity of the tank, not including the volume displaced by the tank.

1. When storing normally stable liquids in vertical cone roof tanks constructed with weak roof-to-shell seam or approved floating roof tanks or when storing crude petroleum in producing areas in any type of tank, one subdivision for each tank in excess of 10,000 bbls. and one subdivision for each group of tanks (no tank exceeding 10,000 bbls. capacity) having an aggregate capacity not exceeding 15,000 bbls.

2. When storing normally stable liquids in tanks not covered in subsection (1), one subdivision for each tank in excess of 100,000 gal [378,500 L] (2,500 bbls.) and one subdivision for each group of tanks (no tank exceeding 100,000 gal (378,500 L) capacity) having an aggregate capacity not exceeding 150,000 gal (567,750 L) (3,570 bbls.).

3. When storing unstable liquids in any type of tank, one subdivision for each tank except that tanks installed in accordance with the drainage requirements of NFPA 15, *Standard for Water Spray Fixed Systems for Fire Protection*, shall require no additional subdivision. Since unstable liquids will react more rapidly when heated than when at ambient temperatures, subdivision by drainage channels is the preferred method.

4. Whenever two or more tanks storing Class I liquids, any one of which is over 150 ft (46 m) in diameter, are located in a common diked area, intermediate dikes shall be provided between adjacent tanks to hold at least 10 percent of the capacity of the tank so enclosed, not including the volume displaced by the tank.

5. The drainage channels or intermediate curbs shall be located between tanks so as to take full advantage of the available space with due regard for the individual tank capacities. Intermediate curbs, where used, shall be not less than 18 in. (457.2 mm) in height.

"With due regard for the individual tank capacities" indicates that an intermediate curb should be further removed from the larger of two adjoining different-sized tanks.

(h) Where provision is made for draining water from diked areas, such drains shall be controlled in a manner so as to prevent flammable or combustible liquids from entering natural water courses, public sewers, or public drains, if their presence would constitute a hazard. Control of drainage shall be accessible under fire conditions from outside the dike.

Paragraph (h) requires that where provision is made for draining water from dike areas, controls shall be accessible outside the diked area and shall prevent flammable or combustible liquids from escaping. This is because uncontrolled escape from a dike area would defeat the purpose of the dike itself.

(i) Storage of combustible materials, empty or full drums, or barrels shall not be permitted within the diked area.

Paragraph (i) prohibits storage of combustible materials or empty or full drums within a dike area. The presence of such material might cause, or contribute to, a fire, and uses up some of the required volumetric capacity of the dike enclosure.

2-2.4 Normal Venting for Aboveground Tanks.

This section considers venting only from the fire protection standpoint. Greater strictures may be imposed by the Environmental Protection Agency. When such strictures involve the manifolding of vents, reference should be made to 1-1.2. Also, because flame arresters are not effective in long vent lines, lines may need to be designed to withstand the pressures imposed by compression ignition ("dieseling"). *Inerting* of the manifolded system or other means may be necessary to avoid hazardous situations. See NFPA 69.[7]

2-2.4.1 Atmospheric storage tanks shall be adequately vented to prevent the development of vacuum or pressure sufficient to distort the roof of a cone roof tank or exceeding the design pressure in the case of other atmospheric tanks, as a result of filling or emptying, and atmospheric temperature changes.

The prohibition against distorting the roof of a cone roof tank applies specifically to tanks built in accordance with API Standard 650, Section 2-1.3.1(b). Roof distortion should, by design, cause failure of the weak roof-to-shell type of construction. This should occur only under emergency conditions. (See Figure 2-12.)

Figure 2-12. Tanks having inadequate normal venting capacity. Every atmospheric storage tank must be equipped with normal vents of sufficient size that will prevent the development of an internal vacuum or pressure sufficient to distort the roof of a cone roof tank, or exceed the design pressure in other than atmospheric tanks.

TANK STORAGE 55

The prohibition against exceeding the design pressure applies only to tanks listed in 2-1.4.2(a) and (b) since these are the only ones built to standards that specify a design pressure and require testing to assure that the specification has been met.

Tanks built in accordance with UL 142, Section 2-1.3.1(a), are not built to any design pressure. Horizontal tanks are required to pass a leakage test using at least 5 psig air pressure. Vertical tanks, when placed in installed position, may also be leak-tested to at least 5 psig air pressure or tested by filling with water and then applying air. However, vertical tanks may be built with a weak seam roof, provided they are 10 ft (3 m) or more in diameter. The latter tanks are leak tested only to 1½ psig. All types of UL tanks are arbitrarily considered to have a design pressure of 1 psig. (See 2-1.4.3 and Figure 2-13.)

In addition to being properly designed to withstand pressures produced by the stored liquid, a tank must also be adequately vented to permit filling and emptying.

In order to fill a tank, air and vapor must get out, or the tank will become pressurized. Pushing out this air and vapor requires that the pressure in the tank be slightly above atmospheric pressure. For this reason, tanks are designed to withstand an internal pressure of 8 in. (203.2 mm) of water gage (⅓ psi) (2.298 kPa).

The emptying of a tank results in air coming in. If this were not the case, the tank would become underpressurized. To allow air to get in, the pressure in the tank must be slightly below atmospheric pressure. For this reason, tanks are designed to withstand a tank vacuum of 2.5 in. (63.5 mm) of water gage (¹⁄₁₀ psi) (0.689 kPa).

Considering the low pressure for which a tank is designed might lead one to question why more tanks don't fail, even given safety factors of 2 to 1 or higher. Adequate venting is the answer. If the vent is large enough to relieve all of the pressure imposed, and is kept clear at all times, tanks will neither implode nor explode.

2-2.4.2 Normal vents shall be either sized in accordance with: (1) the American Petroleum Institute Standard No. 2000, *Venting Atmospheric and Low-Pressure Storage Tanks, 1973**, or (2) other accepted standard; or shall be at least as large as the filling or withdrawal connection, whichever is larger, but in no case less than 1¼ in. (31.75 mm) nominal inside diameter.

Table 2-2 presents Table I from API Standard No. 2000, 1973, titled "Thermal Venting Capacity Requirements."[8] The notes following Table I from API 2000, 1973, are important in explaining the philosophy behind the figures. Consequently, "(2) other accepted standard" is included to cover cases where it is desired to make special calculations to cover special conditions (such as off-gassing, when crude oil is delivered from an oil field) or extreme temperature changes (as from night to day) in desert country; (3) is applicable

---
*Available from American Petroleum Institute, 2101 L St., N.W., Washington, DC 20037.

56 FLAMMABLE AND COMBUSTIBLE LIQUIDS CODE HANDBOOK

Figure 2-13. A Simple Graphic Demonstration of Atmospheric Tank Strength

A STORAGE TANK IS DESIGNED:

1. TO HOLD LIQUID

Liquid exerts pressure on the sides and base of the tank.
Pressure = heat of liquid.

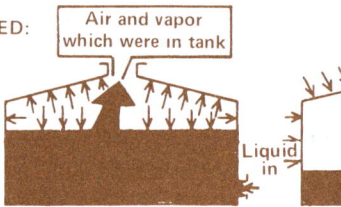

2. TO BE FILLED (Air and vapor which were in tank)

For liquid to get in, air and vapor must get out. If they can't, the tank will be pressurized. For air and vapor to be pushed out, the pressure in the tank must be slightly above atmospheric pressure. The tank is designed for an internal pressure of 8 in. water gage (WG).

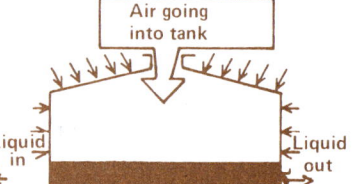

3. TO BE EMPTIED (Air going into tank)

For liquid to get out, air must get in. If it can't, the tank will be underpressured. For air to be sucked in, the pressure in the tank must be slightly below atmospheric pressure.
The tank is designed for an external pressure (or vacuum in the tank) of 2 $\frac{1}{2}$ in. WG.

WHAT ARE INCHES WATER GAGE?

They are a measurement of pressure, used for very low pressures:
8 in. WG = $\frac{1}{3}$ lb/sq in.
2 $\frac{1}{2}$ in. WG = $\frac{1}{10}$ lb/sq in.

Or put another way:

2 $\frac{1}{2}$ in. WG is the pressure at the bottom of a cup of tea.

8 in. WG is the pressure at the bottom of a pint of beer.

YOU CAN BLOW OR SUCK ABOUT 24 IN. WG

24 in.

24 in.

That means by just using your lungs you could overpressure or underpressure a storage tank.
(Because of the volume of air, it would take you a long time).

If you don't believe it, because storage tanks always look big and strong, just study the following table.

If a baked bean tin has a strength of 1, then:

	SHELL	ROOF
Baked bean tin (small)	1	1
40 gal drum	$\frac{1}{2}$	$\frac{1}{3}$
50 m^3 tank	$\frac{1}{3}$	$\frac{1}{8}$
100 m^3 tank	$\frac{1}{4}$	$\frac{1}{11}$
500 m^3 tank	$\frac{1}{6}$	$\frac{1}{33}$
1,000 m^3 tank	$\frac{1}{8}$	$\frac{1}{57}$

Next time you eat baked beans, just see how easy it is to push the sides or top in with your fingers — and then look at the table again. (Any small tin will do if you don't like baked beans.)

Note too: The bigger the tank, the more fragile it is. The roof is weaker than the shell.

Up to 1,000 m^3, the tank shell and roof are only as thick as the line under these words.

TANK STORAGE

IF ALL THAT'S TRUE, IS A STORAGE TANK STRONG ENOUGH?

Yes. A 1,000 m^3 tank has factor of safety of 2 against failure (smaller tanks have bigger safety factors) — provided it is operated within the very low pressures allowed.

Most of the pressures we have available are *many times bigger* than the allowable pressures; that is, 8 in. WG inside, 2 $\frac{1}{2}$ in. WG outside.

FOR EXAMPLE:

Full atmospheric pressure outside	= 150 times bigger
Transfer pump head inside	= 120 times bigger
40 psi nitrogen inside	= 120 times bigger
100 psi steam inside	= 300 times bigger

All of these pressures or even a small part of them will cause the tank to

IMPLODE   EXPLODE

HOW DO WE STOP THIS HAPPENING?

By making sure that:
1. The tank has a vent big enough to relieve all sources of pressure that might be applied to it.
2. The vent is always clear.
3. The vent is never modified without the authorization of the plant or section engineer.

Here are some typical faults in vents which should never happen

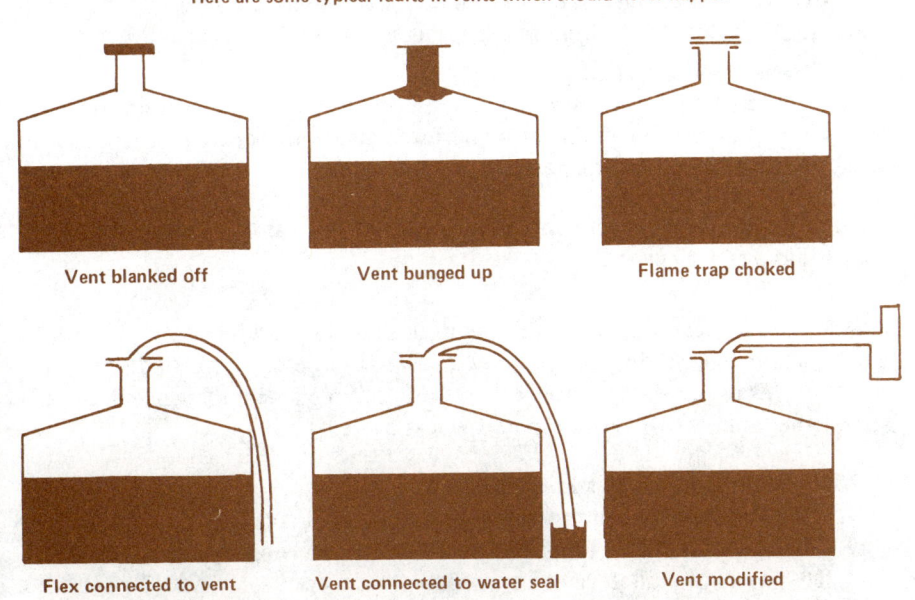

Vent blanked off   Vent bunged up   Flame trap choked

Flex connected to vent   Vent connected to water seal   Vent modified

Adapted with permission from Safety Newsletter No. 115 of the Imperial Chemical Industries, Ltd.

mostly to small tanks which might be filled or emptied rapidly. It should be remembered that API cone roof tanks will fail at the weak seam if overfilled. This problem cannot be solved by the vapor vents provided in the roof. An overflow line, not covered in the *Code*, is often used for such protection. Since the overflow line may contain liquid or other obstructions, it cannot be counted on to furnish part of the required venting capacity.

2-2.4.3 Low-pressure tanks and pressure vessels shall be adequately vented to prevent development of pressure or vacuum, as a result of filling or emptying and atmospheric temperature changes, from exceeding the design pressure of the tank or vessel. Protection shall also be provided to prevent overpressure from any pump discharging into the tank or vessel when the pump discharge pressure can exceed the design pressure of the tank or vessel.

Up to pressures of 1 lb (6.89 kPa) or so, pallet type (weighted check) valves can be used to prevent overpressure. Many of these valves also incorporate a vacuum breaker. For tanks built to API 620, a conventional relief valve may be used. In that case a separate vacuum-breaking device may be needed unless the vapor pressure of the stored product, under all conditions of storage, is high enough to prevent developing a dangerous vacuum under conditions of maximum possible withdrawal rate. For tanks built to UL 142, the combination device is usually used. Tanks built to the ASME *Code for Unfired Pressure Vessels* [2-1.4.2(b)], are often built to withstand full vacuum, so they may not need a vacuum breaker. While an open vent may be adequate to protect the tank from pressure rupture or vacuum collapse, an open vent must not be used if the tank contains a Class I liquid, and should not be used for any liquid heated above its flash point except as permitted in 2-2.4.6. (See 1-1.3.)

2-2.4.4 If any tank or pressure vessel has more than one fill or withdrawal connection and simultaneous filling or withdrawal can be made, the vent size shall be based on the maximum anticipated simultaneous flow.

The word "tank" covers both atmospheric tanks and low-pressure tanks, so all situations are covered.

2-2.4.5 The outlet of all vents and vent drains on tanks equipped with venting to permit pressures exceeding 2.5 psig (17.2 kPa) shall be arranged to discharge in such a way as to prevent localized overheating of, or flame impingement on, any part of the tank, in the event vapors from such vents are ignited.

This requirement was added after a fire (one of several similar fires) suddenly escalated at the Shamrock Oil Company in the Texas Panhandle on July 19, 1956. A spheroid of 15,000 barrel (630,000 gal) capacity contained about 12,000 barrels of a mixture of pentane (boiling point 97°F [36.1°C]) and hexane (boiling point 156.2°F [69°C]). The Reid vapor pressure was 10.2 psia.

TANK STORAGE 59

Figure 2-14. Table I from API Standard No. 2000, 1973: Thermal Venting Requirements (Expressed in cu ft of free air per hr — 14.7 psia at 60°F [15.6°C]). *

Tank Capacity**		Inbreathing (Vacuum) All Stocks	Outbreathing (Pressure) Flash Point 100°F (37.8°C) or Above	Flash Point Below 100°F (37.8°C)
(Barrels) 1	(Gallons)	2	3	4
60	2,500	60	40	60
100	4,200	100	60	100
500	21,000	500	300	500
1,000	42,000	1,000	600	1,000
2,000	84,000	2,000	1,200	2,000
3,000	126,000	3,000	1,800	3,000
4,000	168,000	4,000	2,400	4,000
5,000	210,000	5,000	3,000	5,000
10,000	420,000	10,000	6,000	10,000
15,000	630,000	15,000	9,000	15,000
20,000	840,000	20,000	12,000	20,000
25,000	1,050,000	24,000	15,000	24,000
30,000	---	28,000	17,000	28,000
35,000	---	31,000	19,000	31,000
40,000	---	34,000	21,000	34,000
45,000	---	37,000	23,000	37,000
50,000	---	40,000	24,000	40,000
60,000	---	44,000	27,000	44,000
70,000	---	48,000	29,000	48,000
80,000	---	52,000	31,000	52,000
90,000	---	56,000	34,000	56,000
100,000	---	60,000	36,000	60,000
120,000	---	68,000	41,000	68,000
140,000	---	75,000	45,000	75,000
160,000	---	82,000	50,000	82,000
180,000	---	90,000	54,000	90,000

**Interpolate for intermediate sizes.

NOTES:

1. For tanks with a capacity of more than 20,000 barrels (840,000 gal), the requirements for the vacuum condition are very close to the theoretically computed value of 2 cu ft of air per hr per sq ft of total shell and roof area.

2. For tanks with a capacity of less than 20,000 barrels (840,000 gal), the thermal inbreathing requirements for the vacuum condition have been based on 1 cu ft of free air per hr for each barrel of tank capacity. This is substantially equivalent to a mean rate of vapor space-temperature change of 100°F (37.8°C) per hour.

3. For stocks with a flash point of 100°F (37.8°C) or above, the outbreathing requirement has been assumed at 60 percent of the inbreathing capacity requirement. The tank roof and shell temperatures cannot rise as rapidly under any condition as they can drop, such as during a sudden cold rain.

4. For stocks with a flash point below 100°F (37.8°C), the thermal pressure-venting requirement has been assumed equal to the vacuum requirement in order to allow for vaporization at the liquid surface and for the higher specific gravity of the tank vapors.

*Table I, API Standard No. 2000, Second Edition, December 1973, "Venting Atmospheric and Low-Pressure Storage Tanks" is reproduced with permission of American Petroleum Institute, 2101 L Street, Northwest, Washington, DC 20037.

The Reid vapor pressure is measured with the liquid at a temperature of 100°F (37.8°C), at which temperature 14.7 psia Reid is the boiling point. The spheroid was designed for 15 psig, in accordance with API 620. The spheroid

was exposed to a liquid ground fire that was being fought by the plant fire brigade with assistance from others. The contents soon reached the boiling point. The vapors coming from the vent caught fire. The vent had a weather hood which directed the flame down onto the top of the tank. Since the tank was not full, the top was not cooled by its contents. The heated tank top lost strength and the tank ruptured. The resultant fire ball killed 19 fire fighters. (See Figure 2-15.)

Figure 2-15. Nineteen fire fighters lost their lives when a 15,000-barrel spheroid containing a 12,000-barrel mixture of pentane and hexane was exposed to a liquid fire being fought by a fire brigade at the Shamrock Oil Company in the Texas Panhandle in 1956. When the contents of the spheroid reached the boiling point, the vapors coming from the vent caught fire. The vent had a weather hood which directed the flame down onto the top of the tank and since the tank was not full, the top was not cooled by its contents. The heated top lost strength, failed, and the resulting fire ball killed the 19 fire fighters. This fire, among several others, resulted in the addition of 2-2.4.5 to NFPA 30.

If vents are equipped with outlet pipes that lack weather hoods, rain water or condensate can collect and freeze. Thus, open weep holes are provided to drain off water. Flammables emitted from these holes when the vent operates must be able to burn without heating the top of a tank and pressurizing the tank to more than 2.5 psig. Failure of a tank top at a pressure less than 2.5 psig will be gradual and will not create a hazardous fireball.

2-2.4.6 Tanks and pressure vessels storing Class IA liquids shall be equipped with venting devices which shall be normally closed except when venting to pressure or vacuum conditions. Tanks and pressure vessels storing Class IB and IC liquids shall be equipped with venting devices which shall be normally closed except when venting under pressure or vacuum conditions, or with listed flame arresters. Tanks of 3,000 bbls. capacity or less containing crude petroleum in crude-producing areas, and outside aboveground atmospheric tanks under 1,000 gal (3,785 L) capacity containing other than Class 1A liquids may have open vents. (*See 2-2.6.2.*)

> These requirements are intended to limit the escape of hazardous quantities of flammable vapors in the case of Class IA liquids, which boil below 100°F (37.8°C). In an open tank, such liquids will always be at a temperature below ambient because they are cooled by the loss of heat of vaporization required to produce the vapors needed to fill the vapor space of the tank. When an open vent makes vapor space unlimited, significant generation of vapors is continuous.
>
> In Class IB and Class IC liquids there is the potential, not present in the Class IA liquids, that the vapor space may be in the flammable range. Therefore, flame arresters are permitted as an alternate to keeping the vents closed, except when actually venting. Because of their higher boiling points and accompanying lower vapor pressures, the degree of liquid subcooling caused by vaporization, and hence the rate of vaporization, is insignificant from a flammability standpoint.
>
> The exemption of relatively small crude gathering tanks occurs because such tanks are generally in sparsely populated areas and usually have vapor spaces too rich to burn by reason of release of dissolved gases.
>
> The exemption of tanks under 1,000 gal (3,785 L) capacity containing other than Class IA liquids is due to the relatively insignificant rate of vapor release. In addition, the vent location requirement given in 2-2.6.2 makes it highly improbable that a continuous vapor trail in the flammable range will exist between the vapor space of the tank and any exterior source of ignition.

2-2.4.7 Flame arresters or venting devices required in 2-2.4.6 may be omitted for IB and IC liquids where conditions are such that their use may, in case of obstruction, result in tank damage. Liquid properties justifying the omission of such devices include, but are not limited to, condensation, corrosiveness, crystallization, polymerization, freezing or plugging. When any of these conditions exist, consideration may be given to heating, use of devices employing special materials of construction, the use of liquid seals, or inerting (*see* NFPA 69, *Standard on Explosion Prevention Systems*).

An alternate not mentioned here is frequent inspection and periodic removal of the arresting element for shop cleaning, while replacing the element with a spare. If this procedure is followed, care must be exercised to avoid mechanical damage to the arresting element and to see that it is properly installed.

2-2.5 Emergency Relief Venting for Fire Exposure for Aboveground Tanks.

When exposed to fire, the liquid content of a tank will be heated and may boil, producing evolution of vapor in excess of that described in 2-2.4.1 for normal operating conditions. (See Figure 2-16.) Provisions for safely releasing this vapor are described in the following paragraphs.

Figure 2-16. (Left) Failure to provide adequate emergency relief venting can cause tanks to explode violently, or (Right) to rocket great distances.

2-2.5.1 Except as provided in 2-2.5.2, every aboveground storage tank shall have some form of construction or device that will relieve excessive internal pressure caused by exposure fires.

This section was drastically revised, and Appendix A was added to the *Code*, in the 1963 edition. The emergency venting section of API Standard No. 2000 (see 2-2.4.2 where the normal venting requirements are given) was revised in 1966 so that the requirements given here and in API 2000 are identical. Prior to 1963, this *Code* used suggested emergency vent sizes based on a heat input of 6,000 Btu/hr/sq ft of tank surface (A) wetted by the contents. Prior to 1966, API Standard No. 2000 used 21,000 $A^{0.82}$. Experimental work reported in 1942-43 by Duggan, Gilmour, and Fisher suggested use of a figure of 18,000 to 24,000 A, with no exponent. A subcommittee labored from 1943 to 1963 to try to find a consensus that would eliminate the conflict. (See Appendix A.)

TANK STORAGE 63

The consensus was reached following a fire in Kansas City, Kansas, on August 18, 1959, during which five fire fighters and one spectator were killed when a 21,000-gal (79 485-L) horizontal tank of gasoline failed and the tank rocketed. (See Figure 2-17.) Two brick walls were knocked down by the flying tank which, after traveling 94 ft (28.6 m), landed 15 ft (4.5 m) into the street where most of the fire fighters were on hose lines. The tank failed because of internal overpressure and softening of the metal at the end remote from the fire fighters. A fire engine was burned up.

Figure 2-17. Five fire fighters and 1 spectator were killed and 64 fire fighters injured when a ball-of-fire issued from a "skyrocketing" 21,000-gal (79 485-L) tank during a bulk plant fire in Kansas City, Kansas. Two brick walls were knocked down by the flying tank which traveled 94 ft (28.6 m) from its concrete supports landing 15 ft (4.5 m) into the street where most of the fire fighters were on hose lines. The inadequately vented tank failed from built-up pressure after approximately 1½ hr fire exposure. (Photo Courtesy of UPI.)

2-2.5.2 Tanks larger than 12,000 gal (45 420 L) capacity storing Class IIIB liquids and not within the diked area or the drainage path of Class I or Class II liquids do not require emergency relief venting.

Class IIIB liquids have low vapor pressures and high boiling points. In the time required to start boiling and thus cause excessive internal pressure, tanks of this size will have metal above the internal liquid level so heated and

64   FLAMMABLE AND COMBUSTIBLE LIQUIDS CODE HANDBOOK

softened that the metal will fail and the tank will be self-venting. However, if the tank is in an area containing a still-burning liquid, it then might add fuel to that fire.

2-2.5.3 In a vertical tank the construction referred to in 2-2.5.1 may take the form of a floating roof, lifter roof, a weak roof-to-shell seam, or other approved pressure relieving construction. The weak roof-to-shell seam shall be constructed to fail preferential to any other seam.

In a fire emergency, the design of a weak roof-to-shell seam allows the roof to tear free from the shell as the pressure builds up. The tank is then fully vented. Without this emergency relief feature, the tank might rupture randomly, or even rocket, as may happen when the bottom-to-shell seam fails first. (See Figure 2-18.)

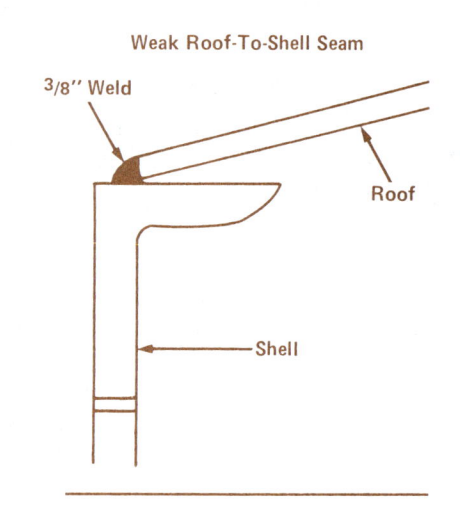

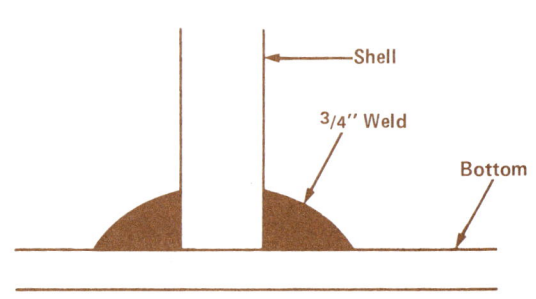

Figure 2-18. Vertical cone roof tanks built either to the specification of Underwriters Laboratories Inc. or the American Petroleum Institute may have a weak roof-to-shell seam. The roof is attached to the shell of the tank, usually on an angle iron, and only a ⅜-in. (.95-cm) weld on one side of the roof-to-shell joint is used. On the bottom-to-shell seam, both sides of the joint are secured by a weld of at least ¾ in. (1.9 cm) — making the bottom-to-shell seam much stronger than the top-to-shell seam. Thus, any pressure that develops during fire exposure conditions to such a tank will be relieved by the roof tearing loose from the shell on part of the tank. In some cases, an internal explosion of the vapor space will completely blow the roof from the tank. In such a case, the entire roof serves as an explosion vent.

2-2.5.4 Where entire dependence for emergency relief is placed upon pressure relieving devices, the total venting capacity of both normal and emergency vents shall be enough to prevent rupture of the shell or bottom of the tank if vertical, or of the shell or heads if horizontal. If unstable liquids are stored, the effects of heat or gas resulting from polymerization, decomposition, condensation, or self-reactivity shall be taken into account. The total capacity of both normal and emergency venting devices shall be not less than that derived from Table 2-8 except as provided in 2-2.5.6 or 2-2.5.7. Such device may be a self-closing manhole cover, or one using long bolts that permit the cover to lift under internal pressure, or an additional or larger relief valve or valves. The wetted area of the tank shall be calculated on the basis of 55 percent of the total exposed area of a sphere or spheroid, 75 percent of the total exposed area of a horizontal tank and the first 30 ft (9.1 m) abovegrade of the exposed shell area of a vertical tank. (*See Appendix A for the square footage of typical tank sizes.*)

> The reference to unstable liquids (see definition in Chapter 1) is intentionally vague. The heat released by a chemical reaction in the contents of a tank will often be many times greater than the heat input from a surrounding fire. Styrene and methyl acrylate are examples of such contents. However, the latter releases heat more rapidly than the former. Sizing of emergency vents for tanks containing unstable liquids should be done on a case-by-case basis by someone familiar with the thermodynamics of the specific heat-releasing reaction. Another implication is that evolution of vapor that must be vented will, if the liquid is stable, take place only at the portions of the container in contact with the liquid. When the liquid is unstable, vapor or gas evolution will also take place throughout the reacting contents. In this latter case, the liquid swells because of the bubbles of gas or vapor in all its portions, rather than in the relatively small volume in contact with the heated shell. This usually means that a two-phase mixture of liquid and vapor is discharged from the vent, so the vent size must be large enough to compensate for the drop in capacity occasioned by this mixed flow as opposed to the flow of all vapor. Vent sizes here are calculated for stable liquids and vapor venting.
>
> The figures prescribed for determining how much of the total surface of a tank shall be considered wetted are based on a consensus among Technical Committee members who had witnessed fire tests and accidental fires and their results. They also considered the "view factor"; i.e., how much energy in the flame is seen by the tank, and the improbability of a tank being completely surrounded by an "optically thick flame." Such a flame is one that is thick enough [about 15 ft (4.5 m)] so that energy radiated into the tank cannot reradiate through the flame and be dissipated into the environment.
>
> Tables 2-8 and 2-9 are expressed in cubic feet of free air per hour, rather than as size of opening in the tank. This is because any venting device, even a manhole opening, has a specific discharge coefficient. Complicated devices, such as flame arresters, have coefficients that vary widely according to design. Also, flow through any specific device of a specific design would be much

Table 2-8
Wetted Area Versus Cubic Feet Free Air per Hour*
(14.7 psia and 60° F) (101.3 kPa and 15.6° C)

Sq. Ft.	CFH	Sq. Ft.	CFH	Sq. Ft.	CFH
20	21,100	200	211,000	1,000	524,000
30	31,600	250	239,000	1,200	557,000
40	42,100	300	265,000	1,400	587,000
50	52,700	350	288,000	1,600	614,000
60	63,200	400	312,000	1,800	639,000
70	73,700	500	354,000	2,000	662,000
80	84,200	600	392,000	2,400	704,000
90	94,800	700	428,000	2,800	742,000
100	105,000	800	462,000	and over	
120	126,000	900	493,000		
140	147,000	1,000	524,000		
160	168,000				
180	190,000				
200	211,000				

SI Units: 1 sq ft = 0.0929 sq m; 1 cu ft = 0.02832 cu m.
*Interpolate for intermediate values.

Table 2-9
Wetted Area Over 2,800 sq ft and Pressures Over 1 psig

Sq. Ft.	CFH	Sq. Ft.	CFH
2,800	742,000	9,000	1,930,000
3,000	786,000	10,000	2,110,000
3,500	892,000	15,000	2,940,000
4,000	995,000	20,000	3,720,000
4,500	1,100,000	25,000	4,470,000
5,000	1,250,000	30,000	5,190,000
6,000	1,390,000	35,000	5,900,000
7,000	1,570,000	40,000	6,570,000
8,000	1,760,000		

SI Units: 1 sq ft = 0.0929 sq m; 1 cu ft = 0.02832 cu m.

greater when venting from a 5 lb (34.7 kPa) tank, down to atmospheric, than when venting from an atmospheric tank where the permissible pressure drop through the device might be only a few inches of water pressure. The tables are derived from the material given in Appendix A. Appendix A should be consulted for a further discussion of the emergency venting recommendations.

2-2.5.5 For tanks and storage vessels designed for pressures over 1 psig (6.895 kPa), the total rate of venting shall be determined in accordance with Table 2-8, except that when the exposed wetted area of the surface is greater than 2,800 sq ft (260 m²), the total rate of venting shall be in accordance with Table 2-9 or calculated by the following formula:

$$\text{CFH} = 1{,}107 \, A^{0.82}$$

Where:

CFH = venting requirement, in cubic feet of free air per hour

A = exposed wetted surface, in square feet

The foregoing formula is based on $Q = 21{,}000 \, A^{0.82}$.

> As pointed out in Appendix A, pressure vessels or low-pressure tanks will ordinarily be used for the storage of materials with low boiling points. Consequently, some emergency venting is needed to keep the tank from failing at a pressure greatly in excess of the relief valve setting. Following the venting suggestions in Table 2-9 will alleviate this problem. The equation CFH = $1{,}107 A^{0.82}$ is derived by substituting $21{,}000 \, A^{0.82}$ for Q in the formula in Appendix A, CFH = $\dfrac{70.5 Q}{L\sqrt{M}}$ where $L\sqrt{M}$ is the figure for n-Hexane, 1337.

2-2.5.6 The total emergency relief venting capacity for any specific stable liquid can be determined by the following formula:

$$\text{Cubic feet of free air per hour} = V \, \frac{1{,}337}{L\sqrt{M}}$$

V = cubic feet of free air per hour from Table 2-8

L = latent heat of vaporization of specific liquid in Btu per pound

M = molecular weight of specific liquids

> For many materials, notably alcohols, $L\sqrt{M}$ is larger than 1,337. This paragraph permits smaller emergency vents for tanks storing such materials. However, tanks may later be used for materials other than those for which they were originally intended, so caution should be used when accepting a vent having less capacity than would be required for n-Hexane.

2-2.5.7 For tanks containing stable liquids, the required airflow rate of 2-2.5.4 or 2-2.5.6 may be multiplied by the appropriate factor listed in the following schedule when protection is provided as indicated. Only one factor can be used for any one tank.

.5 for drainage in accordance with 2-2.3.2 for tanks over 200 sq ft (18.6 m²) of wetted area

.3 for water spray in accordance with NFPA 15, *Standard for Water Spray Fixed Systems for Fire Protection*, and drainage in accordance with 2-2.3.2

.3 for insulation in accordance with 2-2.5.7(a)

.15 for water spray with insulation in accordance with 2-2.5.7(a) and drainage in accordance with 2-2.3.2 (*see Appendix A*)

(a) Insulation systems for which credit is taken shall meet the following performance criteria:

1. Remain in place under fire exposure conditions.

2. Withstand dislodgment when subjected to hose stream impingement during fire exposure. This requirement may be waived where use of solid hose streams is not contemplated or would not be practical.

3. Maintain a maximum conductance value of 4.0 Btu's per hour per square foot per degree F (Btu/hr/sq ft/ °F) when the outer insulation jacket or cover is at a temperature of 1,660 °F (904.4 °C) and when the mean temperature of the insulation is 1,000 °F (537.8 °C).

The justification for the factors by which the venting capacity may be reduced is:

1. Paragraph 2.2.3.2 requires that no spill fire can seriously expose a tank. A 50 percent factor is reasonable.

2. Experimental work, summarized in A-4.4.3.2 of NFPA 15,[9] shows that heat input to protected tanks is reduced to about 6,000 Btu/hr/sq ft of surface exposed to fire and wetted by the contents, and 6,000 divided by 20,000 is 30 percent. However, this factor should not be used unless the water supply to the system is adequate to keep it operating, bearing in mind other demands on the water supply, until the emergency requiring use of the water spray system is over. Some tank farm fires have lasted for days. It must also be considered that water spray systems are subject to internal plugging and damage by explosion.

3. The insulation requirements are not so specific as they seem. Paragraph (a)(1) requires that it remain in place under fire exposure conditions, although neither this nor the other requirements insist that it be undamaged after the fire and hose stream exposure. It is enough that it remain functional while it is being exposed to fire. For example, foam glass and hydrous calcium silicate held on by stainless steel bands can give protection for many hours, particularly if put on in multilayers with the joints broken. They may need replace-

ment when the fire is out. Only materials such as mineral wool and refractory materials are likely to survive a severe fire without need for replacement.

There is a good discussion of insulation systems in the National Academy of Sciences publication *Pressure-Relieving Systems for Marine Cargo Bulk Liquid Containers.*[10] There is some thought that "an emergency is an emergency is an emergency," and that emergency vent sizes should never be reduced, regardless of the permission to do so granted by this *Code*.

2-2.5.8 The outlet of all vents and vent drains on tanks equipped with emergency venting to permit pressures exceeding 2.5 psig (17.2 kPa) shall be arranged to discharge in such a way as to prevent localized overheating of or flame impingement on any part of the tank, in the event vapors from such vents are ignited.

The reason for this was discussed under 2-2.4.5.

2-2.5.9 Each commercial tank venting device shall have stamped on it the opening pressure, the pressure at which the valve reaches the full open position and the flow capacity at the latter pressure. If the start to open pressure is less than 2.5 psig (17.2 kPa) and the pressure at full open position is greater than 2.5 psig (17.2 kPa), the flow capacity at 2.5 psig (17.2 kPa) shall also be stamped on the venting device. The flow capacity shall be expressed in cubic feet per hour of air at 60°F (15.6°C) and 14.7psia (101.3 kPa).

(a) The flow capacity of tank venting devices under 8 in. (203.2 mm) in nominal pipe size shall be determined by actual test of each type and size of vent. These flow tests may be conducted by the manufacturer if certified by a qualified impartial observer, or may be conducted by a qualified, impartial outside agency. The flow capacity of tank venting devices 8 in. (203.2 mm) nominal pipe size and larger, including manhole covers with long bolts or equivalent, may be calculated provided that the opening pressure is actually measured, the rating pressure and corresponding free orifice area are stated, the word "calculated" appears on the nameplate, and the computation is based on a flow coefficient of 0.5 applied to the rated orifice area.

(b) A suitable formula for this calculation is:

$$CFH = 1,667 \, C_f \, A \, \sqrt{P_t - P_a}$$

where CFH = venting requirement in cubic feet of free air per hour
C_f = 0.5 [the flow coefficient].
A = the orifice area in sq in.
P_t = the absolute pressure inside the tank in inches of water.
P_a = the absolute atmospheric pressure outside the tank in inches of water.

The requirement that the venting capacity be stamped on the device is to eliminate the need to consult vendors' catalogs. This eases the inspector's job.

The formula given in (b) uses $C_f = 0.5$ as an approximation for the true coefficient which would be made up of a basic orifice factor, a viscosity factor based on the Reynolds number of the vapor, an expansion factor because the vapor expands as it leaves the orifice, a temperature factor, and a specific gravity factor. The use of 0.5 results in a conservative estimate that will predict a somewhat lower flow than is actually the case, and hence gives a small factor of safety in some cases.

2-2.6 Vent Piping for Aboveground Tanks.

2-2.6.1 Vent piping shall be constructed in accordance with Chapter 3.

There is a slight ambiguity here since 3-1.1 refers to piping containing liquids. However, ANSI B31[11] covers piping for gases under vacuum as well as liquids, so the reference to ANSI B31 in 3-1.1 is pertinent here.

2-2.6.2 Where vent pipe outlets for tanks storing Class I liquids are adjacent to buildings or public ways, they shall be located so that the vapors are released at a safe point outside of buildings and not less than 12 ft (3.7 m) above the adjacent ground level. In order to aid their dispersion, vapors shall be discharged upward or horizontally away from closely adjacent walls. Vent outlets shall be located so that flammable vapors will not be trapped by eaves or other obstructions and shall be at least 5 ft (1.5 m) from building openings.

This *Code* relates only to fire and explosion hazards. The vent termination locations permitted by this paragraph may not be suitable if the vapors are also toxic or have an objectionable odor.

The 5 ft (1.5 m) required from building openings and the 12 ft (3.7 m) above ground level are based on an engineering estimate of the distance an ignitable concentration of vapors may exist around the end of a vent pipe. If the hazard extends 5 ft (1.5 m) down, its edge is 12 minus 5 or 7 ft (2.1 m) above grade so that ignition by a smoker, walking or driving by, would not occur. The size of the hazardous volume may be estimated by taking the case of hexane. A sphere of 5 ft (1.5 m) radius has a volume of 523.6 cu ft (14.8 m³). The lower flammable limit is about 1 percent and would require 5.236 cu ft (0.157 m³) of hexane vapor. At 60 °F (15.6 °C), hexane has a vapor pressure of 100 mm of mercury (13.3 kPa), so the vapors coming out of the vent would be 100/760 or 13 percent hexane — too rich to burn. These vapors would produce 5.236/0.13 or 40 cu ft (1.13 m³) of vapor, more than enough to fill the 5 ft (1.5 m) sphere with a mixture at the lower flammable limit. If such vapor quantities were discharged into air moving at 1 mph, this would constitute 88 ft/min (1609 m/s). (This is below 1 on the Beaufort Scale, at which point the leaves on trees do not move and smoke goes straight up.) The vapors would be diluted out of the hazardous range long before they filled a 5 ft (1.5 m) radius sphere. The conclusion is that the 5 ft (1.5 m) rule gives ample protection.

When vent piping is modified by the addition of items such as devices to absorb or adsorb unwanted components from the stream being vented, these devices should not unduly restrict the flow of vapors. Also, the piping should not be modified in such a way that liquid collecting in low points could either permit the tank to be overpressured or start a siphoning action that could implode the tank.

2-2.6.3 The manifolding of tank vent piping shall be avoided except where required for special purposes such as vapor recovery, vapor conservation or air pollution control. When tank vent piping is manifolded, pipe sizes shall be such as to discharge, within the pressure limitations of the system, the vapors they may be required to handle when manifolded tanks are subject to the same fire exposure.

The manifolding of vents creates a hazardous situation if the vent system may, even infrequently, contain a mixture in the flammable range. There are no listed flame arresters that will prevent flame propagation through piping unless they are installed within a maximum distance, specified in the listing, from the open ends of the pipes. This listing assumes that the ignition will take place at the open end, where the pressure would initially be atmospheric. In a manifolded piping system there would be no open end.

So-called "flame checks," used in piping systems feeding premixed air and gas into the burners in an oven or furnace, are usually of small diameter and act to prevent flame travel in only one direction, from the furnace back into the piping.

The manifolding of vents from vessels operating under pressure can involve the possibility of diesel-type ignition of flammable mixtures in "dead ends" of the system.

2-2.6.4 Vent piping for tanks storing Class I liquids shall not be manifolded with vent piping for tanks storing Class II or Class III liquids unless positive means are provided to prevent the vapors from Class I liquids from entering tanks storing Class II or Class III liquids, to prevent contamination (*see 1-1.2*) and possible change in classification of the less volatile liquid.

A practical way to prevent such contamination is to maintain a low flow of inert gas from the vent of each tank containing a Class II or Class III liquid. Air should not be used as a purge.

2-2.7 Tank Openings Other Than Vents for Aboveground Tanks.

2-2.7.1 Each connection to an aboveground tank through which liquid can normally flow shall be provided with an internal or an external valve located as close as practical to the shell of the tank.

Because repair or maintenance of an internal valve usually requires that the tank be emptied, internal valves are seldom used on larger tanks when only flammable or combustible liquids are involved.

2-2.7.2 Each connection below the liquid level through which liquid does not normally flow shall be provided with a liquidtight closure. This may be a valve, plug or blind, or a combination of these.

Since a valve can be opened by mistake, resulting in a spill of tank contents, it is advisable to plug or blank the valve outlet when a valve not intended for operating purposes is installed.

2-2.7.3 Openings for gaging on tanks storing Class I liquids shall be provided with a vaportight cap or cover. Such covers shall be closed when not gaging.

Gaging openings may also be used for taking samples. Metallic cups, floats, and the like should have electrical continuity with the tank shell to avoid the possibility of ignition by static spark, and provision should be made for a 30 minute static dissipation period before gaging or sampling after filling a tank. Although the atmosphere in a tank containing a Class I liquid is normally too rich to burn, vapors escaping from the gaging opening form a hazardous zone in the immediate vicinity. If the tank is being emptied or recently has had liquid withdrawn from it, much of the vapor space in the tank may be in the flammable range.

2-2.7.4 For Class IB and Class IC liquids other than crude oils, gasolines and asphalts, the fill pipe shall be so designed and installed as to minimize the possibility of generating static electricity. A fill pipe entering the top of a tank shall terminate within 6 in. (152.4 mm) of the bottom of the tank and shall be installed to avoid excessive vibration.

This is the first place in the *Code* that static electricity has been mentioned as a possible source of ignition. The exceptions in the first sentence recognize that: (1) crude oils and asphalt have such a low electrical resistivity that static generation has not proved to be a problem, and (2) gasoline normally has a vapor-air mixture at its surface too rich to be ignitable, so that static ignition is not possible. The manner in which liquid is admitted to a tank can cause turbulence, particularly if the inlet pipe terminates above the liquid surface. Therefore, it is required that the pipe terminate close to the bottom. (See Figure 2-19.) A similar warning is contained in 6-3.7.4. See also NFPA 77.

This paragraph addresses only one way to avoid generating a static charge. There are many ways to charge a fluid. Charged mists result when a stream of falling liquid breaks up into droplets. Flow of liquid through a pipe charges the liquid; this effect is greatly enhanced if the flow involves two or more phases. The voltage generated may be so high that the charge leaks off to

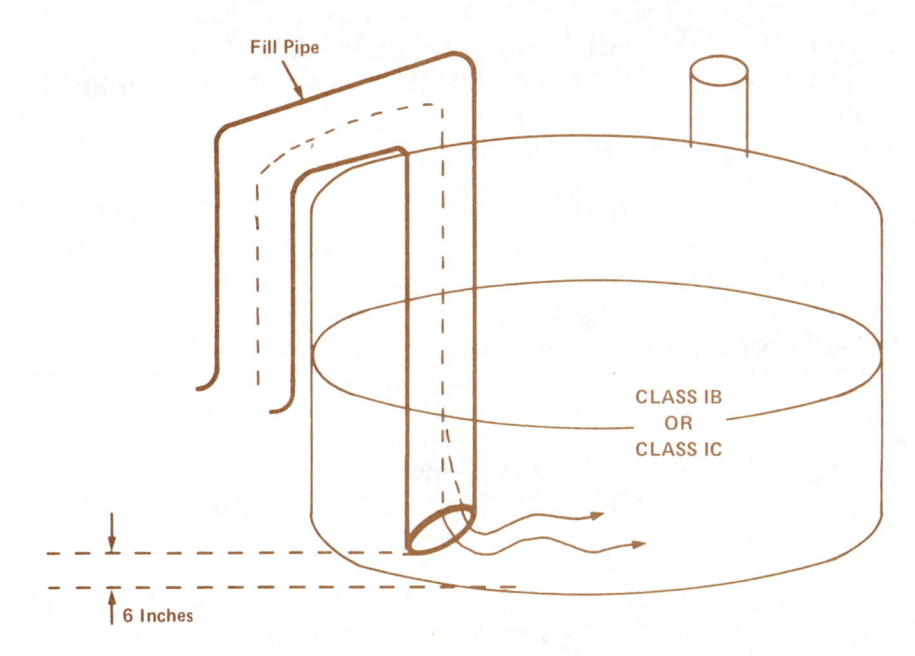

Figure 2-19. A fill pipe entering the top of a tank should terminate within 6 in. (152.4 mm) of the bottom of the tank and should be installed to avoid excessive vibration.

grounded objects by way of a nonincendive corona discharge. The generation of voltages of this magnitude is not possible with liquids having a resistance of less than 10^{10} ohm-centimeters because the opposite charges generated by fluid motion recombine too rapidly. If the fluids have high resistivity, the charge is not mobile and may discharge to a grounded object. Although the voltage difference is high, the discharge has insufficient energy to be incendive. The danger of producing an incendive static spark exists when a charged fluid of high resistance flows over or near an ungrounded conductor. This conductor will reach the voltage of the surrounding fluid. If the ungrounded charged conductor is then near a grounded conductor, an incendive spark can jump because the charge on the charged conductor is mobile. Accordingly, no ungrounded conductor should exist within, or be introduced into, a system that might become charged. Ungrounded conductors may be obvious, such as metal floats on level gages, or unobvious, such as slugs of conducting fluid, metal parts of intrinsically safe electrical equipment, or metal sampling cups introduced by use of a wood rod or a nylon cord.

The reference to vibration in the last sentence of 2-2.7.4 is a warning that a pipe supported only at the top might vibrate and break off, without anyone becoming aware of it. The resulting falling stream would void the purpose of having it extend to within 6 in. (152.4 mm) of the bottom.

NFPA 77[12] gives much additional information.

2-2.7.5 Filling and emptying connections for Class I, Class II and Class IIIA liquids which are made and broken shall be located outside of buildings at a location free from any source of ignition and not less than 5 ft (1.5 m) away from any building opening. Such connections for any liquid shall be closed and liquidtight when not in use and shall be properly identified.

2-3 Installation of Underground Tanks.

2-3.1 Location. Excavation for underground storage tanks shall be made with due care to avoid undermining of foundations of existing structures. Underground tanks or tanks under buildings shall be so located with respect to existing building foundations and supports that the loads carried by the latter cannot be transmitted to the tank. The distance from any part of a tank storing Class I liquids to the nearest wall of any basement or pit shall be not less than 1 ft (0.3048 m), and to any property line that can be built upon, not less than 3 ft (0.91 m). The distance from any part of a tank storing Class II or Class III liquids to the nearest wall of any basement, pit or property line shall be not less than 1 ft (0.3048 m).

> The first requirement in restricting the location of underground tanks is that the excavation shall not jeopardize the foundation of structures, or be so placed that settlement of the latter can damage the tank. The minimum spacing of 1 ft (0.3048 m) from the tank to the wall of a basement or pit is to provide space for protective measures, such as excavation, well points, etc. If a leak should develop, in the case of Class I liquids, it could result in the presence of flammable vapors in the adjacent underground space.
>
> The distance requirements give the minimum required for installation purposes. With Class I liquids, the 3 ft (0.91 m) distance to a property line that may be built upon is required to minimize the possibility of damage to a gasoline tank or to its protective sand or gravel envelope by construction activities on adjacent property. (See Figure 2-20.)

2-3.2 Burial Depth and Cover.

2-3.2.1 Steel underground tanks shall be set on firm foundations and surrounded with at least 6 in. (152.4 mm) of noncorrosive inert material such as clean sand or gravel well-tamped in place. The tank shall be placed in the hole with care, since dropping or rolling the tank into the hole can break a weld, puncture or damage the tank, or scrape off the protective coating of coated tanks.

> Back filling with gravel or sand containing large stones can likewise damage a tank's coating or even cause dents or gouges. Gravel should be in accordance with the dictionary definition; i.e., "consist of rounded pebbles." Because it may contain corrosive salt, sea sand should not be used.

2-3.2.2 Steel underground tanks shall be covered with a minimum of 2 ft (.6096 m) of earth, or shall be covered with not less than 1 ft (0.3048 m) of earth, on top of which shall be placed a slab of reinforced concrete not less than 4 in. (101.6

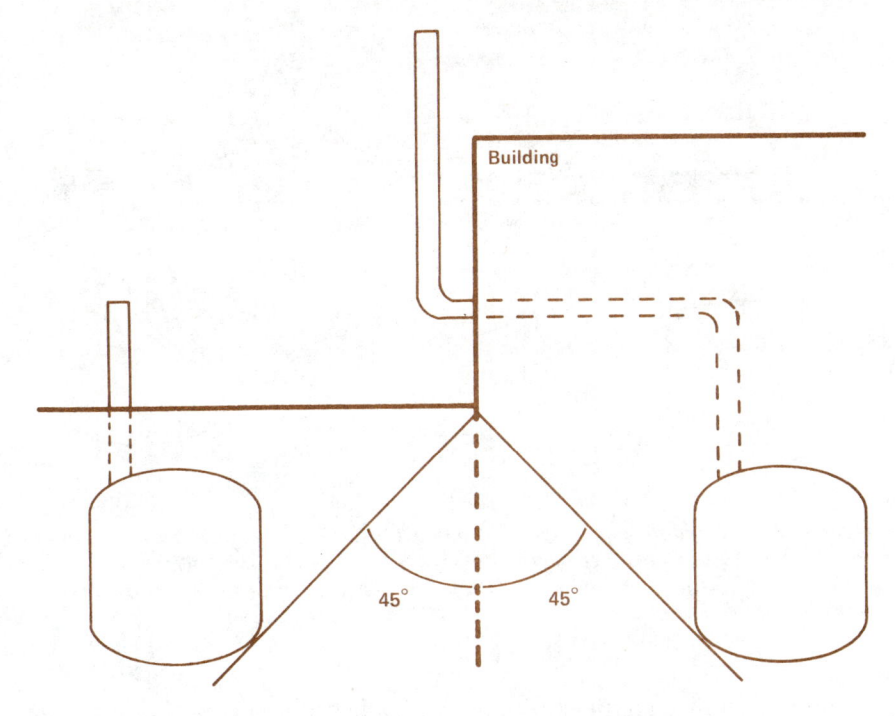

Figure 2-20. A common rule of thumb used by many engineers when locating a tank below the bottom foundation of a building is to place the tank outside of these 45-degree angles measured from the extension of the foundation. This rule of thumb holds true for tanks located under a building or just outside a building.

mm) thick. When they are, or are likely to be, subjected to traffic, they shall be protected against damage from vehicles passing over them by at least 3 ft (0.91 m) feet of earth cover, or 18 in. (457.2 mm) of well-tamped earth plus either 6 in. (152.4 mm) of reinforced concrete or 8 in. (203.2 mm) of asphaltic concrete. When asphaltic or reinforced concrete paving is used as part of the protection, it shall extend at least 1 ft (0.3048 m) horizontally beyond the outline of the tank in all directions.

See Figure 2-21.

2-3.2.3 Nonmetallic underground tanks shall be installed in accordance with the manufacturer's instructions. The minimum depth of cover shall be as specified in 2-3.2.2 for steel tanks.

The specified methods of covering are intended to be adequate to prevent damage to the tank if a loaded truck is driven over the spot where it is buried.

2-3.2.4 For tanks built in accordance with 2-1.3.1(a), the burial depth shall be such that the static head imposed at the bottom of the tank will not exceed 10 psig

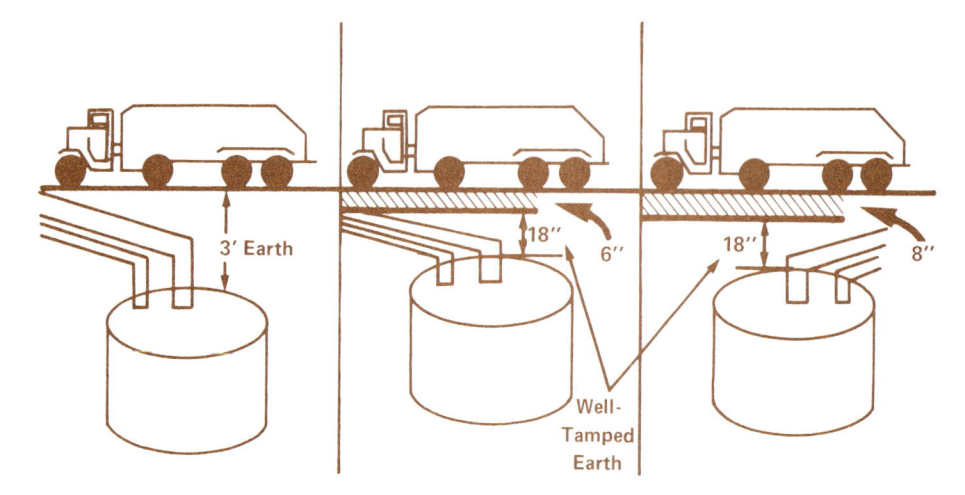

Figure 2-21. When steel underground tanks are located in an area likely to be subjected to traffic, they must be protected against damage from vehicles passing over them by at least 3 ft (0.91 m) of earth cover, or 18 in. (457.2 mm) of well-tamped earth plus either 6 in. (152.4 mm) of reinforced concrete or 8 in. (203.2 mm) of asphaltic concrete.

if the fill or vent pipe are filled with liquid. If the depth of cover is greater than the tank diameter, the tank manufacturer shall be consulted to determine if reinforcement is required.

The limit of 10 psig at the bottom of the tank means that if the tank and vent pipe were filled with water, the maximum permissible distance from the bottom of the tank to the top of the vent would be 10/0.434 or 23 ft (7 m). Assuming they were filled with gasoline, the permissible height would be (for 0.7 specific gravity gasoline) 23/.7, or about 33 ft (10 m). However, the specific gravity of other liquids may be higher than gasoline.

The second sentence expresses concern that excessive external loading may cause partial tank collapse with consequent sudden ejection of part of the contents.

2-3.3 Corrosion Protection. Unless tests show that soil resistivity is 10,000 ohm-centimeters or more, and there are no other corrosive conditions, tanks and their piping shall be protected by either:

(a) a properly installed and maintained cathodic protection system with or without coatings, or

(b) corrosion resistant materials of construction such as special alloys, fiber glass reinforced plastic, or fiber glass reinforced plastic coatings, or equivalent approved system. Selection of the type of protection to be employed shall be based upon the corrosion history of the area and the judgment of a qualified engineer.

(*See API Publication 1615-1979, Installation of Underground Petroleum Storage Systems, for further information.**)

It is highly important that buried tanks remain leaktight during all of their service life. It is unrealistic to require maximum corrosion protection for tanks in all locations. In favorable soils tanks have lasted leak-free for 20 years and longer. In other locations, tanks have leaked from corrosion within a very few years after installation. An engineering study is necessary to evaluate the rate of corrosion and the consequences if leaks should occur, taking into account both the economic loss and the possible effect on others if leaking liquid should spread off the owner's property. NFPA 329[13] discusses the source and control of such leaks. Also, see API 841-41490.

2-3.4 Abandoned Underground Tanks.

2-3.4.1 Underground tanks taken out of service shall be safeguarded or disposed of in a safe manner. (*See Appendix B.*)

See Figure 2-22.

Figure 2-22. (Left) Before. (Right) After. Failure to follow the procedures described in 2-3.4.1 has resulted in major problems and even fatalities. For example, in the case shown here, a tank had been abandoned underground with gasoline still in it. During the spring, the high water table caused the gasoline from the abandoned tank to enter a crawl space underneath a service station building. Three fire fighters and 2 service station employees were killed when an explosion occurred in the crawl space.

2-3.5 Vents for Underground Tanks.

The treatment of vents in this section differs slightly from that in 2-2.4. The differences arise for several reasons: buried tanks never have a cone roof or a

---
*Available from American Petroleum Institute, 2101 L St., N.W., Washington, DC 20037.

floating roof; they are generally of the horizontal cylindrical type; they cannot be seen when being filled or emptied; and the contents are never heated by the sun or severely chilled in cold weather.

2-3.5.1 Location and Arrangement of Vents for Class I Liquids. Vent pipes from underground storage tanks storing Class I liquids shall be so located that the discharge point is outside of buildings, higher than the fill pipe opening, and not less than 12 ft (3.6 m) above the adjacent ground level. Vent pipes shall not be obstructed by devices provided for vapor recovery or other purposes unless the tank and associated piping and equipment are otherwise protected to limit backpressure development to less than the maximum working pressure of the tank and equipment by the provision of pressure-vacuum vents, rupture discs or other tank venting devices installed in the tank vent lines. Vent outlets and devices shall be protected to minimize the possibility of blockage from weather, dirt or insect nests, and shall be so located and directed that flammable vapors will not accumulate or travel to an unsafe location, enter building openings or be trapped under eaves. Tanks containing Class IA liquids shall be equipped with pressure and vacuum venting devices which shall be normally closed except when venting under pressure or vacuum conditions. Tanks storing Class IB or Class IC liquids shall be equipped with pressure-vacuum vents or with listed flame arresters. Tanks storing gasoline are exempt from the requirements for pressure and vacuum venting devices, except as required to prevent excessive back pressure, or flame arresters, provided the vent does not exceed 3 in. (76.20 mm) nominal inside diameter. (*See also 7-2.1.1.*)

In some industrial installations where numerous horizontal tanks have been grouped in one area and either buried or mounded over, vents terminating less than 12 ft (3.6 m) above ground level have been accepted if the area is fenced to limit access and the termination is above normal snow level.

Use of the term "maximum working pressure" instead of "design pressure" makes it possible, in the opinion of the Technical Committee, to accept 2 ½ psig as the maximum working pressure for a tank built to UL 58. This is one half of the minimum pressure required in the test for tightness before installation. If the tank is built to ASME *Code for Unfired Pressure Vessels*, the design working pressure is given in the papers that accompany the tank. Since leakage from underground tanks is not apparent and usually depends on inventory control for detection (see 7-2.1.4), there is a detailed requirement to prevent overpressure.

The permission to use an open vent pipe of 3 in. (76 mm) inside diameter or smaller, provided it terminates 12 ft (3.6 m) aboveground and discharges upward (see 7-2.1.1), is based on the properties of gasoline vapor. The properties of gasoline are intentionally varied so that it has a lower initial boiling point in cold weather, and so that the vapor pressure at 100°F (37.8°C) is higher in winter than in summer. On the other hand, the initial boiling point and vapor pressure at 100°F (37.8°C) are lower when the gasoline is to be used at high altitudes, such as in Colorado rather than in New England. The specifications to ensure that it has the proper qualities are given in ASTM D-439, *Standard Specifications for Automotive Gasoline.*[15] Bureau of Mines Bulletin 627, *Flammability Characteristics of Combustible Gases and Vapors,*[16] indicates that at 70°F (21.1°C), for example, a gasoline-air mixture in equilibrium with liquid gasoline would contain 25 percent gasoline vapor. Since the flammable range (again in accordance with Bureau of Mines tests) is 1.4 to 7.6, the vapors in a buried tank would be too rich to burn. The vapors would need to be diluted with more than three times their volume of air to drop into the flammable range. Because of the changes in composition to meet conditions of use, as required by ASTM D-439, the vapors coming out of a storage tank that is being filled will always be too rich to burn. If a fire were lit at the outlet of the vent pipe while the buried tank was being filled, it could not flash back because the vapor in the vent line would be too rich to burn. If a fire at the vent were burning at the moment of starting to pump gasoline out of the tank, the flame would simply be sucked down into the pipe and be snuffed out because of lack of oxygen.

2-3.5.2 Vent Capacity. Tank venting systems shall be provided with sufficient capacity to prevent blowback of vapor or liquid at the fill opening while the tank is being filled. Vent pipes shall not be less than 1¼ in. (31.75 mm) nominal inside diameter. The required venting capacity depends upon the filling or withdrawal rate, whichever is greater, and the vent line length. Unrestricted vent piping sized in accordance with Table 2-10 will prevent back-pressure development in tanks from exceeding 2.5 psig (17.24 kPa). Where tank venting devices are installed in vent lines, their flow capacities shall be determined in accordance with 2-2.5.9.

Table 2-10
Vent Line Diameters

Maximum Flow GPM	50 ft.	Pipe Length* 100 ft.	200 ft.
100	1¼-inch	1¼-inch	1¼-inch
200	1¼-inch	1¼-inch	1¼-inch
300	1¼-inch	1¼-inch	1½-inch
400	1¼-inch	1½-inch	2-inch
500	1½-inch	1½-inch	2-inch
600	1½-inch	2-inch	2-inch
700	2-inch	2-inch	2-inch
800	2-inch	2-inch	3-inch
900	2-inch	2-inch	3-inch
1000	2-inch	2-inch	3-inch

SI Units: 1 in. = 25.40 mm; 1 ft = 0.3048 m.
*Vent lines of 50 ft., 100 ft. and 200 ft. of pipe plus 7 ells.

80   FLAMMABLE AND COMBUSTIBLE LIQUIDS CODE HANDBOOK

The preceding paragraph recognizes that vent lines may run a considerable distance, usually below ground, before rising up to the termination point. Piping larger than that in Table 2-10 will usually be needed if a flame arrester or pressure-vacuum vent is installed, as would be the case with many Class IB and IC liquids.

2-3.5.3 Location and Arrangement of Vents for Class II or Class IIIA Liquids. Vent pipes from tanks storing Class II or Class IIIA liquids shall terminate outside of building and higher than the fill pipe opening. Vent outlets shall be above normal snow level. They may be fitted with return bends, coarse screens or other devices to minimize ingress of foreign material.

By omission, vents for Class IIIB liquids may terminate indoors. However, this may not apply to heated Class IIIB liquids. (See 1-1.3.)

2-3.5.4 Vent piping shall be constructed in accordance with Chapter 3. Tank vent pipes and vapor return piping shall be installed without sags or traps in which liquid can collect. Condensate tanks, if utilized, shall be installed and maintained so as to preclude the blocking of the vapor return piping by liquid. The vent pipes and condensate tanks shall be located so that they will not be subjected to physical damage. The tank end of the vent pipe shall enter the tank through the top.

As noted under 2-2.6.2, vent piping is not really addressed in Chapter 3.

2-3.5.5 When tank vent piping is manifolded, pipe sizes shall be such as to discharge, within the pressure limitations of the system, the vapors they can be required to handle when manifolded tanks are filled simultaneously. Float-type check valves installed in tank openings connected to manifolded vent piping to prevent product contamination may be used provided that the tank pressure will not exceed that permitted by 2-3.2.4 when the valves close.

Exception: For service stations, the capacity of manifolded vent piping shall be sufficient to discharge vapors generated when two manifolded tanks are simultaneously filled.

The exception for service stations is made because delivery trucks are not currently equipped to fill more than two tanks simultaneously.

2-3.5.6 Vent piping for tanks storing Class I liquids shall not be manifolded with vent piping for tanks storing Class II or Class III liquids unless positive means are provided to prevent the vapors from Class I liquids from entering tanks storing Class II or Class III liquids, to prevent contamination (*see 1-1.2*) and possible change in classification of the less volatile liquid.

Since underground openings will be connected to piping, 3-7.1 should be consulted for test methods.

2-3.6 Tank Openings Other Than Vents for Underground Tanks.

2-3.6.1 Connections for all tank openings shall be liquidtight.

2-3.6.2 Openings for manual gaging, if independent of the fill pipe, shall be provided with a liquidtight cap or cover. Covers shall be kept closed when not gaging. If inside a building, each such opening shall be protected against liquid overflow and possible vapor release by means of a spring loaded check valve or other approved device.

> Note that 2-3.1 does not prohibit having a tank under a building on the same premises, provided that the restrictions with respect to walls, foundations, and supports are met. However, the only permitted indoor access to the tank contents is via the specially equipped gaging opening.

2-3.6.3 Fill and discharge lines shall enter tanks only through the top. Fill lines shall be sloped toward the tank. Underground tanks for Class I liquids having a capacity of more than 1,000 gal (3,785 L) shall be equipped with a tight fill device for connecting the fill hose to the tank.

> This is to minimize loss of liquid through piping leaks, since pipes installed as required will contain liquid only when in use. With improperly sloped or bottom connected pipe, liquid would always be present.

2-3.6.4 For Class IB and Class IC liquids other than crude oils, gasolines and asphalts, the fill pipe shall be so designed and installed as to minimize the possibility of generating static electricity by terminating within 6 in. (152.4 mm) of the bottom of the tank.

> This is the same as 2-2.7.4. The same comments about static electricity apply.

2-3.6.5 Filling and emptying and vapor recovery connections for Class I, Class II or Class IIIA liquids which are made and broken shall be located outside of buildings at a location free from any source of ignition and not less than 5 ft (1.5 m) away from any building opening. Such connections shall be closed and liquidtight when not in use and shall be properly identified.

> Just as vents from tanks containing Class IIIB liquids are permitted to terminate indoors, filling connections are permitted indoors. As with vents, 1-1.3 is pertinent.

2-3.6.6 Tank openings provided for purposes of vapor recovery shall be protected against possible vapor release by means of a spring-loaded check valve or dry-break connection, or other approved device, unless the opening is pipe-connected to a vapor processing system. Openings designed for combined fill and vapor recovery shall also be protected against vapor release unless connection of the liquid delivery line to the fill pipe simultaneously connects the vapor recovery line. All connections shall be vaportight.

This paragraph is intended to regulate two types of systems. In one, vapor displaced when a storage tank is being filled is piped back to the vapor space of the unit from which the liquid is discharged. This is a vapor recovery system. The opening must not permit vapor to escape from the buried tank except when it is going to the filling unit. In the second type, the vapor is connected to a special unit where it is incinerated, absorbed in oil or other liquid, or adsorbed on activated carbon or similar material for later recovery or disposal by other environmentally acceptable means.

2-4 Installation of Tanks Inside of Buildings.

2-4.1 Location. Tanks shall not be permitted inside of buildings except as provided in Chapters 5, 6, 7, 8 or 9.

As used here, "tank" means storage tank and does not apply to portable tanks, process tanks, etc. Other storage tanks for Class I liquids are prohibited indoors in Chapter 6, Bulk Plants, and are allowed indoors in Chapter 7, Service Stations, only if in special enclosures.

2-4.2 Vents. Vents for tanks inside of buildings shall be as required in 2-2.4, 2-2.5, 2-2.6.2 and 2-3.5, except that emergency venting by the use of weak roof seams on tanks shall not be permitted. Automatic sprinkler systems designed in accordance with the requirements of NFPA 13, *Standard for the Installation of Sprinkler Systems*, may be accepted by the authority having jurisdiction as equivalent to water spray systems for purposes of calculating the required air flow rates for emergency vents in 2-2.5.7. Except for tanks containing Class IIIB liquids, vents shall terminate outside the buildings.

The prohibition against tanks having weak seam roofs should, as a matter of good practice, apply also to tanks with floating roofs. The acceptance of sprinklers as equivalent to directional water spray systems is made on the basis that if the fire is not controlled by sprinklers, structural failure will ruin the tank in any case. Emergency vents are included in the requirement for termination outside the building.

2-4.3 Vent Piping. Vent piping shall be constructed in accordance with Chapter 3.

Here we have the previously mentioned problem (see comments under

2-2.6.1) that Chapter 3, except for its reference to ANSI B31, speaks only to systems containing liquids.

2-4.4 Tank Openings Other Than Vents for Tanks Inside Buildings.

2-4.4.1 Connections for all tank openings shall be liquidtight.

2-4.4.2 Each connection to a tank inside of buildings through which liquid can normally flow shall be provided with an internal or an external valve located as close as practical to the shell of the tank.

Since most leaks occur in piping systems, this requirement minimizes the chance of having an uncontrollable leak.

2-4.4.3 Tanks for storage of Class I or Class II liquids inside buildings shall be provided with either:

(a) a normally closed remotely activated valve,

(b) an automatic-closing heat-activated valve, or

(c) another approved device on each liquid transfer connection below the liquid level, except for connections used for emergency disposal, to provide for quick cut-off of flow in the event of fire in the vicinity of the tank.

This function can be incorporated in the valve required in 2-4.4.2, and if a separate valve, shall be located adjacent to the valve required in 2-4.4.2.

This is to minimize the chance that a fire-related piping failure would cause uncontrollable addition of liquid fuel to the fire.

2-4.4.4 Openings for manual gaging of Class I or Class II liquids, if independent of the fill pipe, shall be provided with a vaportight cap or cover. Openings shall be kept closed when not gaging. Each such opening for any liquid shall be protected against liquid overflow and possible vapor release by means of a spring loaded check valve or other approved device. Substitutes for manual gaging include, but are not limited to, heavy-duty flat gage glasses, magnetic, hydraulic or hydrostatic remote reading devices and sealed float gages.

This expands on the similar requirement for aboveground tanks (2-2.7.3) by suggesting alternates to openings for gaging. Suitable alternates, not mentioned, include mounting the tank on a scale or on load cells.

2-4.4.5 For Class IB and Class IC liquids other than crude oils, gasolines and asphalts, the fill pipe shall be so designed and installed as to minimize the possibility of generating static electricity by terminating within 6 in. (152.4 mm) of the bottom of the tank.

This is the same as 2-2.7.4, and the same comments apply.

2-4.4.6 The fill pipe inside of the tank shall be installed to avoid excessive vibration of the pipe.

This is, in essence, the second sentence of 2-2.7.4.

2-4.4.7 The inlet of the fill pipe and the outlet of a vapor recovery line for which connections are made and broken shall be located outside of buildings at a location free from any source of ignition and not less than 5 ft (1.5 m) away from any building opening. Such connections shall be closed and tight when not in use and shall be properly identified.

The rationale for the 5 ft (1.5 m) spacing was discussed under 2-2.6.2.

2-4.4.8 Tanks storing Class I, Class II and Class IIIA liquids inside buildings shall be equipped with a device, or other means shall be provided, to prevent overflow into the building. Suitable devices include, but are not limited to, a float valve, a preset meter on the fill line, a valve actuated by the weight of the tank contents, a low head pump which is incapable of producing overflow, or a liquid-tight overflow pipe at least one pipe size larger than the fill pipe discharging by gravity back to the outside source of liquid or to an approved location.

Here, Class IIIB liquids are excluded from the requirement. If heated, they may not be excluded. (See 1-1.3.)

2-4.4.9 Tank openings provided for purposes of vapor recovery shall be protected against possible vapor release by means of a spring-loaded check valve or dry-break connections, or other approved device, unless the opening is pipe-connected to a vapor processing system. Openings designed for combined fill and vapor recovery shall also be protected against vapor release unless connection of the liquid delivery line to the fill pipe simultaneously connects the vapor recovery line. All connections shall be vaportight.

This is identical with 2-3.6.6, and the same comments apply.

2-5 Supports, Foundations and Anchorage for All Tank Locations.

2-5.1 Tanks shall rest on the ground or on foundations made of concrete, masonry, piling or steel. Tank foundations shall be designed to minimize the possibility of uneven settling of the tank and to minimize corrosion in any part of the tank resting on the foundation. Appendix E of API Standard 650, *Specification for Welded Steel Tanks for Oil Storage*, and Appendix B of API Standard 620, *Recommended Rules for the Design and Construction of Large, Welded, Low-Pressure Storage Tanks*,* provide information on tank foundations.

2-5.2 When tanks are supported above the foundations, tank supports shall be installed on firm foundations. Supports for tanks storing Class I, Class II or

---

*Available from American Petroleum Institute, 2101 L St., N.W., Washington, DC 20037.

TANK STORAGE

Class IIIA liquids shall be of concrete, masonry or protected steel. Single wood timber supports (not cribbing) laid horizontally may be used for outside aboveground tanks if not more than 12 in. (304.8 mm) high at their lowest point.

Supports of unprotected steel are prohibited. Experience has shown that they soften and fail after brief exposure to a fire in liquid on the ground, resulting in rupture of piping and possible spill of tank contents. (See Figure 2-22.) The exception for solid timber supports less than 12 in. (304.8 mm) high is based on the "slow burning" characteristic of "heavy timber construction." (See NFPA *Fire Protection Handbook.*[17]**)**

2-5.3 Steel supports or exposed piling for tanks storing Class I, Class II or Class IIIA liquids shall be protected by materials having a fire resistance rating of not less than 2 hrs, except that steel saddles need not be protected if less than 12 in. (304.8 mm) high at their lowest point. At the discretion of the authority having jurisdiction, water spray protection in accordance with NFPA 15, *Standard for Water Spray Fixed Systems for Fire Protection*, or NFPA 13, *Standard for the Installation of Sprinkler Systems*, or equivalent may be used.

Figure 2-23. (Left) Aboveground tanks on steel supports. (Right) Aftermath of a fire showing collapse of unprotected steel supports. Unfortunately, many tanks are installed on unprotected steel supports even though this practice has been prohibited since the first edition of the Flammable and Combustible Liquids Code *in 1913. When exposed to fire, unprotected steel supports buckle and fail. The sagging supports allow the tank to tip or fall to the ground, breaking piping connections or even damaging the tank. As a result, spilled tank contents allow the fire to spread beyond control.*

2-5.4 The design of the supporting structure for tanks such as spheres shall require special engineering consideration. Appendix N of the API Standard 620, *Recommended Rules for the Design and Construction of Large, Welded, Low-Pressure Storage Tanks*,* contains information regarding supporting structures.

---

*Available from American Petroleum Institute, 2101 L St., N.W., Washington, DC 20037.

2-5.5 Every tank shall be so supported as to prevent the excessive concentration of loads on the supporting portion of the shell.

2-5.6 Where a tank is located in an area subject to flooding, provisions shall be taken to prevent tanks, either full or empty, from floating during a rise in water level up to the established maximum flood stage.

> In several instances high water level has floated full or partially full tanks from their foundations, resulting in hazardous situations in other areas. The requirements in the paragraphs which follow are intended to prevent such incidents.

2-5.6.1 Aboveground Tanks.

2-5.6.1.1 Each vertical tank shall be located so that its top extends above the maximum flood stage by at least 30 percent of its allowable storage capacity.

> A tank so located will not float if filled with gasoline (specific gravity 0.7). It will float if only partially full. To take care of the possibility that a tank might not be full at the time of a flood, the availability of a water supply to fill the tank in case of a flood emergency is required in 2-5.6.1.4.

2-5.6.1.2 Horizontal tanks located so that more than 70 percent of the tank's storage capacity will be submerged at the established flood stage shall be anchored; attached to a foundation of concrete or of steel and concrete of sufficient weight to provide adequate load for the tank when filled with flammable or combustible liquid and submerged by flood water to the established flood stage; or adequately secured from floating by other means. Tank vents or other openings which are not liquidtight shall be extended above maximum flood stage water level.

> The strict limitation on location in 2-5.6.1.1 is relaxed in the case of horizontal tanks, because these tanks can be more reliably anchored. However, it is not practical to provide anchoring for an empty tank, and provision for water loading is still required.

2-5.6.1.3 A dependable water supply shall be available for filling an empty or partially filled tank, except that where filling the tank with water is impractical or hazardous because of the tank's contents, tanks shall be protected by other means against movement or collapse.

2-5.6.1.4 Spherical or spheroid tanks shall be protected by applicable methods as specified for either vertical or horizontal tanks.

2-5.6.2 Underground Tanks.

> Tanks buried in the earth have been known to float up out of the ground if

exposed to high water table or flood waters. Means for preventing this are prescribed in the following paragraphs.

2-5.6.2.1 At locations where there is an ample and dependable water supply available, underground tanks containing flammable or combustible liquids, so placed that more than 70 percent of their storage capacity will be submerged at the maximum flood stage, shall be so anchored, weighted or secured as to prevent movement when filled or loaded with water and submerged by flood water to the established flood stage. Tank vents or other openings which are not liquidtight shall be extended above maximum flood stage water level.

2-5.6.2.2 At locations where there is no ample and dependable water supply or where filling of underground tanks with water is impractical because of the contents, each tank shall be safeguarded against movement when empty, and submerged by high ground water or flood water by anchoring or by securing by other means. Each such tank shall be so constructed and installed that it will safely resist external pressures if submerged.

Reliable means for utilizing water from the flood itself would be considered acceptable.

2-5.6.3 Water Loading. The filling of a tank to be protected by water loading shall be started as soon as flood waters are predicted to reach a dangerous flood stage. Where independently fueled water pumps are relied upon, sufficient fuel shall be available at all times to permit continuing operations until all tanks are filled. Tank valves shall be closed and locked in closed position when water loading has been completed.

2-5.6.4 Operating Instructions.

2-5.6.4.1 Operating instructions or procedures to be followed in a flood emergency shall be readily available.

2-5.6.4.2 Personnel relied upon to carry out flood emergency procedures shall be informed of the location and operation of valves and other equipment necessary to effect the intent of these requirements.

2-5.7 In areas subject to earthquakes, the tank supports and connections shall be designed to resist damage as a result of such shocks.

2-6 Sources of Ignition.

2-6.1 In locations where flammable vapors may be present, precautions shall be taken to prevent ignition by eliminating or controlling sources of ignition. Sources of ignition may include open flames, lightning, smoking, cutting and welding, hot surfaces, frictional heat, sparks (static, electrical and mechanical), spontaneous ignition, chemical and physical-chemical reactions and radiant heat. NFPA 77, *Recommended Practice on Static Electricity*, and NFPA 78, *Lightning Protection Code*, provide information on such protection.

The "precautions" may take the form of signs prohibiting smoking and the use of open flames, restrictions as to the entry of unauthorized people, restrictions concerning automotive equipment, and the enforcement of "hot work" permits for mechanical work or for cutting, welding, and similar procedures.

2-7 Testing.

2-7.1 All tanks, whether shop-built or field-erected, shall be tested before they are placed in service in accordance with the applicable paragraphs of the Code under which they were built. The ASME Code stamp, API monogram, or the Listing Mark of Underwriters Laboratories Inc. on a tank shall be evidence of compliance with this test. Tanks not marked in accordance with the above Codes shall be tested before they are placed in service in accordance with good engineering principles and reference shall be made to the sections on testing in the Codes listed in 2-1.3.1, 2-1.4.2, or 2-1.5.2.

2-7.2 When the vertical length of the fill and vent pipes is such that when filled with liquid the static head imposed upon the bottom of the tank exceeds 10 lbs per sq in. (68.95 kPa), the tank and related piping shall be tested hydrostatically to a pressure equal to the static head thus imposed. In special cases where the height of the vent above the top of the tank is excessive, the hydrostatic test pressure shall be determined by using recognized engineering practice.

The intent is that the tank must be tested to the pressure that would be developed if the tank were accidentally overfilled, causing liquid to rise into the vent pipe. However, under no circumstances should the pressure exceed the design pressure of the tank. (See Figure 2-24.)

2-7.3 In addition to the test called for in 2-7.1 and 2-7.2, all tanks and connections shall be tested for tightness. Except for underground tanks, this tightness shall be made at operating pressure with air, inert gas or water prior to placing the tank in service. In the case of field-erected tanks the test called for in 2-7.1 or 2-7.2 may be considered to be the test for tank tightness. Underground tanks and piping, before being covered, enclosed, or placed in use, shall be tested for tightness hydrostatically, or with air pressure at not less than 3 lbs per sq in. (20.68 kPa) and not more than 5 lbs per sq in. (34.475 kPa). (*See 3-7.1 for testing pressure piping*.)

Positive means, such as a relief valve or regulator, should be incorporated in the air supply to ensure that the test pressure does not exceed that specified. Air tests should not be made on top of stored product. (See Figure 2-25.) See Chapter IV of NFPA 329.[18]

TANK STORAGE   89

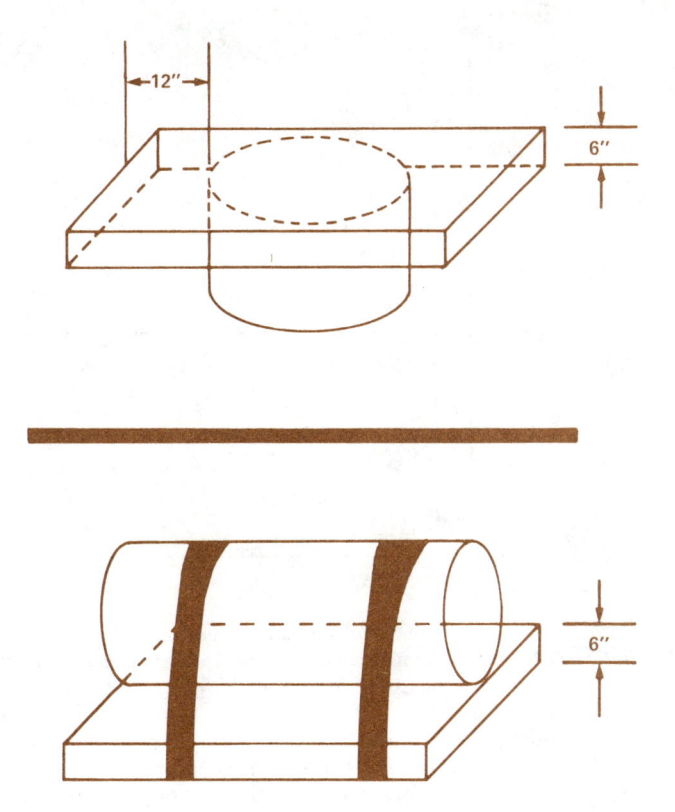

Figure 2-24. One method is to place over the tank 6 in. (152.4 mm) of reinforced concrete extending 12 in. (0.3048 m) beyond the tank in all directions. A second method uses a 6-in. (152.4-mm) reinforced concrete base at the bottom of the hole. Anchoring strips are then placed over the tank and attached to the concrete slab. With both of these methods, the weight of the concrete offsets the buoyancy of the tank.

2-7.4 Before the tank is initially placed in service, all leaks or deformations shall be corrected in an acceptable manner. Mechanical caulking is not permitted for correcting leaks in welded tanks except pin hole leaks in the roof.

2-7.5 Tanks to be operated at pressures below their design pressure may be tested by the applicable provisions of 2-7.1 or 2-7.2 based upon the pressure developed under full emergency venting of the tank.

2-8 Fire Protection and Identification.

2-8.1 A fire extinguishing system in accordance with an applicable NFPA standard shall be provided or be available for vertical atmospheric fixed roof storage tanks larger than 50,000 gal (189,250 L) capacity storing Class I liquids if located in a congested area where there is an unusual exposure hazard to the tank from adjacent property or to adjacent property from the tank. Fixed roof tanks storing

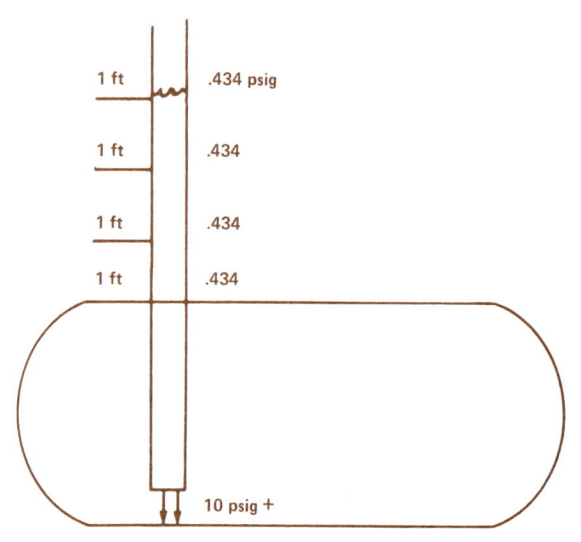

90 FLAMMABLE AND COMBUSTIBLE LIQUIDS CODE HANDBOOK

SI Units: 1 ft = 0.3048 m.

Figure 2-25. When the vertical length of the fill and vent pipes is such that when filled with liquid the static head imposed upon the bottom of the tank exceeds 10 psig, as measured from the bottom of the tank, then the Code requires the tank and related piping to be tested hydrostatically to a pressure equal to the static head thus imposed. When filled with liquid, such pipes can increase the hydrostatic pressure on the bottom of the tank by .434 psig per ft of vertical length above the top of the tank. However, in special cases where the length of the vent above the top of the tank is excessive, the hydrostatic test pressure must be determined by using recognized engineering practice.

Figure 2-26. A typical testing device for determining tank tightness.

Class II or III liquids at temperatures below their flash points and floating roof tanks storing any liquid generally do not require protection when installed in compliance with Section 2-2.

Fire extinguishing systems are not required for small fixed roof tanks because such tanks are seldom involved in fire. If they should become involved there is small risk that the fire will spread. Fire in spills on the ground cannot be extinguished by equipment designed for tank fires, and such fires are usually controlled by public fire departments.

The spacings in Table 2-2 are based on the considerations contained in 2-8.1.

2-8.2 The application of NFPA 704, *Identification of the Fire Hazards of Materials*, to storage tanks containing liquids shall not be required except when the contents have a health or reactivity degree of hazard of 2 or more or a flammability rating of 4. The marking need not be applied directly to the tank but located where it can readily be seen, such as on the shoulder of an accessway or walkway to the tank or tanks or on the piping outside of the diked area. If more than one tank is involved, the markings shall be so located that each tank can readily be identified.

2-9 Prevention of Overfilling of Tanks.

2-9.1 Tanks receiving transfer of Class I liquids from mainline pipelines or marine vessels and located in an area where overfilling may endanger a place of habitation or public assembly shall be either:

(a) Gaged at frequent intervals while receiving transfer of product, and communications maintained with mainline pipeline or marine personnel so that flow can be promptly shut down or diverted, or

(b) Equipped with an independent high level alarm located where personnel are on duty during the transfer and can promptly arrange for flow stoppage or diversion, or

(c) Equipped with an independent high level alarm system that will automatically shut down or divert flow.

The preceding precautions are designed to avoid the risk of fire associated with a spill which can flow liquid to a place where ignition could occur. The requirements are limited to Class I liquids because other classes do not produce vapors subject to ready ignition. Pipeline and marine terminals are singled out for special treatment because the liquid transfer rates are frequently very high, and those receiving the liquid may lack the means or authority to suddenly cut off the flow.

2-10 Leakage Detection and Inventory Records for Underground Tanks.

2-10.1 Accurate inventory records or a leak detection program shall be maintained on all Class I Liquid Storage Tanks for indication of possible leakage from the tanks or associated piping. (*See NFPA 329, Underground Leakage of Flammable and Combustible Liquids*.)

Inventory records have been required by the *Code* prior to this edition, but only in Chapter 7 which governs service stations. The Technical Committee felt that this concept should be applied to all underground tanks containing Class I liquids, and so the preceding paragraph was inserted in Chapter 2.

REFERENCES CITED BY *CODE*

(*These publications comprise a part of the requirements to the extent called for by the* **Code**.)

Underwriters Laboratories Inc., *Standard for Steel Aboveground Tanks for Flammable and Combustible Liquids*, UL 142, Chicago, 1972.

———, *Standard for Steel Underground Tanks for Flammable and Combustible Liquids*, UL 58, Chicago, 1976.

———, *Standard for Steel Inside Tanks for Oil Burner Fuel*, UL 80, Chicago, 1974.

American Petroleum Institute, *Welded Steel Tanks for Oil Storage*, Standard No. 650, Sixth Edition, Washington DC, 1978.

———, *Specification for Bolted Production Tanks*, Standard No. 12B, Twelfth Edition, Washington DC, January 1977.

———, *Specification for Large Welded Production Tanks*, Standard No. 12D, Eighth Edition, Washington DC, January 1977.

———, *Specification for Small Welded Production Tanks*, Standard No. 12F, Seventh Edition, Washington, DC, January 1977.

Steel Tank Institute, Steel Tank Institute Standard STI-P3, *Specifications for STI-P3, System of Corrosion Protection of Underground Steel Storage Tanks*, Chicago, 1980.

American Petroleum Institute, *Recommended Rules for the Design and Construction of Large, Welded, Low-Pressure Storage Tanks*, Standard No. 620, Sixth Edition, Washington DC, 1977.

American Petroleum Institute, *Venting Atmospheric and Low-Pressure Storage Tanks*, Standard No. 2000, Second Edition, Washington DC, 1973.

American Society of Mechanical Engineers, ASME Boiler and Pressure Vessels Code, *Code for Unfired Pressure Vessels*, Section VIII, Division I, New York, 1980.

NFPA 13, *Installation of Sprinkler Systems*, NFPA, Boston, 1978.

NFPA 15, *Standard for Water Spray Fixed Systems for Fire Protection*, NFPA, Boston, 1979.

NFPA 69, *Explosion Prevention Systems*, NFPA, Boston, 1978.

NFPA 329, *Underground Leakage of Flammable and Combustible Liquids*, NFPA, Boston, 1977.

NFPA 77, *Recommended Practice on Static Electricity*, NFPA, Boston, 1977.

NFPA 78, *Lightning Protection Code*, NFPA, Boston, 1977.

NFPA 704, *Identification of the Fire Hazards of Materials*, NFPA, Boston, 1975.

REFERENCES CITED IN COMMENTARY

[1]NFPA 77, *Recommended Practice on Static Electricity*, NFPA, Boston, 1977.

[2]American Petroleum Institute, *Welded Steel Tanks for Oil Storage*, Standard No. 650, Sixth Edition, Washington DC, 1977.

[3] Underwriters Laboratories Inc., *Standard for Steel Aboveground Tanks for Flammable and Combustible Liquids*, UL 142, Chicago, 1972.

[4] American Society of Mechanical Engineers, ASME Boiler and Pressure Vessels Code, *Code for Unfired Pressure Vessels*, Section VIII, Division I, New York, 1980.

[5] American Petroleum Institute, *Guides for Fighting Fires in and around Petroleum Storage Tanks*, API 2021, Washington DC, Nov. 1974.

[6] American Petroleum Institute, *Recommended Rules for the Design and Construction of Large, Welded, Low-Pressure Storage Tanks*, Standard No. 620, Sixth Edition, Washington DC, 1977.

[7] NFPA 69, *Explosion Prevention Systems*, NFPA, Boston, 1978.

[8] American Petroleum Institute, *Venting Atmospheric and Low-Pressure Storage Tanks*, Standard No. 2000, Second Edition, Washington DC, 1973.

[9] NFPA 15, *Standard for Water Spray Fixed Systems for Fire Protection*, NFPA, Boston, 1979.

[10] National Academy of Sciences, *Pressure-Relieving Systems for Marine Cargo Bulk Liquid Containers*, Office of Publications, Washington DC, 1973.

[11] ANSI B31, *American National Standard Code for Pressure Piping*, American National Standards Institute, New York.

[12] NFPA 77, *Recommended Practice on Static Electricity*, NFPA, Boston, 1977.

[13] NFPA 329, *Underground Leakage of Flammable and Combustible Liquids*, NFPA, Boston, 1977.

[14] American Petroleum Institute, *Migration of Petroleum Products in Soil and Ground Water*, API 841-41490, Washington DC, December 1972.

[15] ASTM D-439-79, *Standard Specifications for Automotive Gasoline*, American Society for Testing and Materials, Philadelphia, 1979.

[16] Bureau of Mines, *Flammability Characteristics of Combustible Gases and Vapors*, Bulletin 627, 1965.

[17] McKinnon, G.P., ed., *Fire Protection Handbook*, 14th ed., NFPA, Boston, 1976.

[18] NFPA 329, *Underground Leakage of Flammable and Combustible Liquids*, Chapter IV, NFPA, Boston, 1977.

3

Piping, Valves, and Fittings

Any piping system is susceptible to potential leakage or spills of liquids. Because of the additional fire hazard in piping systems associated with the storage and handling of flammable and combustible liquids, these systems must be able to withstand the intense heat generated by fire for reasonable periods of time while emergency shut-down procedures are implemented and fire fighting measures are commenced. Failure of pipes, valves, and fittings during fire conditions can turn a moderate hazard into an extreme emergency.

3-1 General.

3-1.1 The design, fabrication, assembly, test and inspection of piping systems containing liquids shall be suitable for the expected working pressures and structural stresses. Conformity with the applicable sections of ANSI B31, *American National Standard Code for Pressure Piping*, and the provisions of this chapter shall be considered prima facie evidence of compliance with the foregoing provisions.

It is important to consider the *expected* (or normal) working pressures and structural stresses. This is particularly so when high vapor pressure flammable liquids are involved, or when tanks and piping systems are subject to heating. Heating may result from exposure to direct sunlight or other heating source, from heat radiating off a process or piece of equipment, or from the liquid itself at elevated (heated) temperature.

3-1.2 This chapter does not apply to any of the following:

(a) Tubing or casing on any oil or gas wells and any piping connected directly thereto.

(b) Motor vehicle, aircraft, boat or portable or stationary engine.

(c) Piping within the scope of any applicable boiler and pressure vessel code.

3-1.3 Piping systems consist of pipe, tubing, flanges, bolting, gaskets, valves, fittings, the pressure containing parts of other components such as expansion joints and strainers, and devices which serve such purposes as mixing, separating, snubbing, distributing, metering, or controlling flow.

This paragraph specifies the elements of a piping system as well as its uses.

3-2 Materials for Piping, Valves and Fittings.

3-2.1 Pipe, valves, faucets, fittings and other pressure containing parts as covered in 3-1.3 shall meet the material specifications and pressure and temperature limitations of ANSI B31.3-1973*, *Petroleum Refinery Piping*, or ANSI B31.4-1974*, *Liquid Petroleum Transportation Piping Systems* except as provided by 3-2.2, 3-2.3, and 3-2.4. Plastic or similar materials, as permitted by 3-2.4, shall be designed to specifications embodying recognized engineering principles and shall be compatible with the fluid service.

See Figure 3-1.

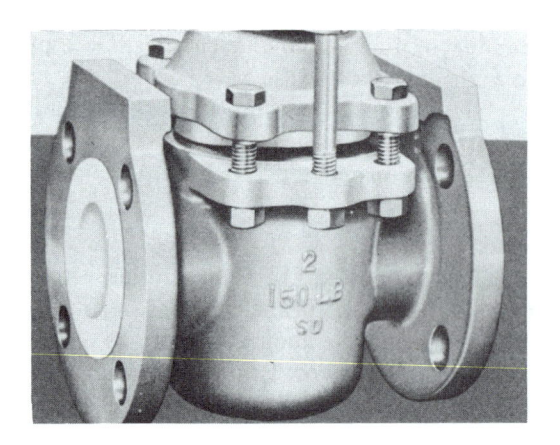

Figure 3-1. Plastic linings can be useful in preventing corrosion in steel pipes. It is for reasons such as this that the Code *allows the use of combustible or noncombustible linings in piping, valves, and fittings.*

3-2.2 Nodular iron shall conform to ASTM A395—74, *Ferritic Ductile Iron Pressure Retaining Castings for Use at Elevated Temperatures.***

3-2.3 Valves at storage tanks, as required by 2-2.7.1 and 2-4.4.2, and their connections to the tank shall be of steel or nodular iron except as provided in 3-2.3.1 or 3-2.3.2.

Valves through which liquid can normally flow, e.g., those on aboveground storage tanks and tanks located inside buildings, shall be of steel or nodular iron except when: (1) the chemical characteristics of the liquid are not compatible with steel (3-2.3.1); (2) the valves are installed internally to the tank

---

*Available from the American Society of Mechanical Engineers, United Engineering Center, 345 East 47th St., New York, NY 10017.

**Available from American Society for Testing and Materials, 1916 Race St., Philadelphia, PA 19103.

(3-2.3.1); (3) crude petroleum tanks meet the requirements of 2-2.2.2 (3-2.3.2); or (4) the tank is used to store Class IIIB liquids and the requirements of 3-2.3.2 are met.

3-2.3.1 Valves at storage tanks may be other than steel or nodular iron when the chemical characteristics of the liquid stored are not compatible with steel or when installed internally to the tank. When installed externally to the tank, the material shall have a ductility and melting point comparable to steel or nodular iron so as to withstand reasonable stresses and temperatures involved in fire exposure, or otherwise be protected such as by materials having a fire resistance rating of not less than 2 hrs.

The reason for requiring steel or nodular iron wherever possible is to ensure that the valve or fittings will be able to withstand reasonable stresses and temperatures created by a fire exposure. An alternative to using steel, nodular iron, or a material similar in properties to steel or nodular iron, is to protect the valve by materials which give a fire resistance rating of 2 hrs.

Internal valves can be of other materials because the liquid keeps the piping and valve cool, thus avoiding failure under fire exposure. (See Figure 3-2.)

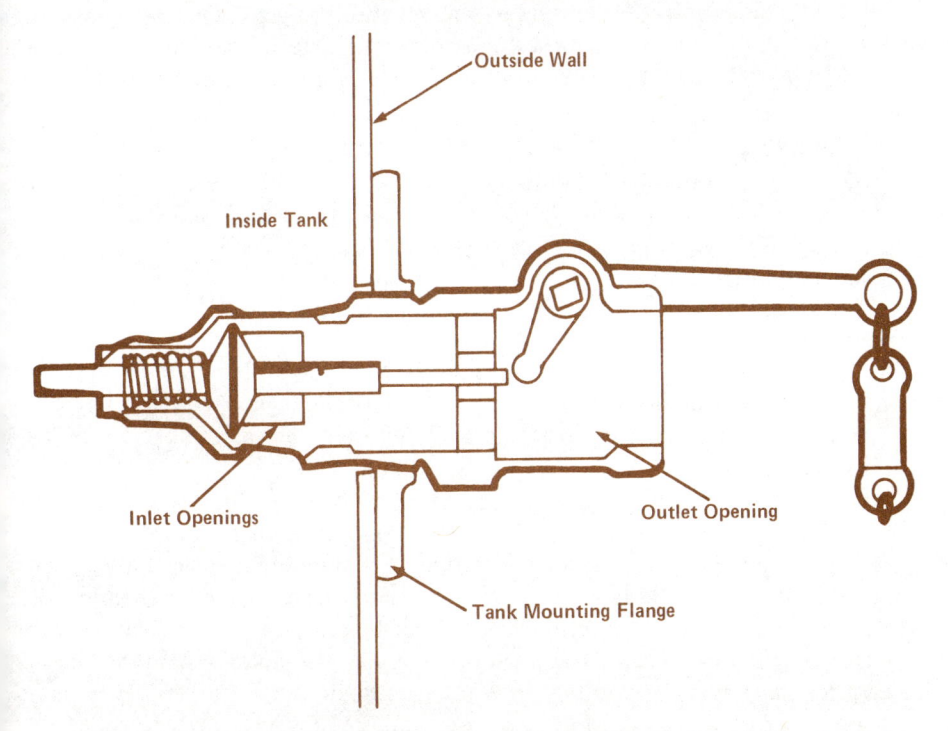

Figure 3-2. The Code permits materials other than steel or nodular iron to be used on valves when installed internally to the tank. The liquid in the tank keeps the piping and valve body material cool and intact so that the contents of the tank would not be released through the piping even under severe fire exposure.

3-2.3.2 Cast iron, brass, copper, aluminum, malleable iron, and similar materials may be used on tanks described in 2-2.2.2 or for tanks storing Class IIIB liquids when the tank is located outdoors and not within a diked area or drainage path of a tank storing a Class I, Class II or Class IIIA liquid.

3-2.4 Low melting point materials, such as aluminum, copper and brass; or materials which soften on fire exposure, such as plastics; or nonductile material, such as cast iron, may be used underground for all liquids within the pressure and temperature limits of ANSI B31, *American National Standard Code for Pressure Piping**. If such materials are used outdoors in aboveground piping systems handling Class I, Class II or Class IIIA liquids or within buildings handling any liquid, they shall be either: (a) suitably protected against fire exposure, or (b) so located that any leakage resulting from the failure would not unduly expose persons, important buildings or structures, or (c) located where leakage can readily be controlled by operation of an accessible remotely located valve or valves.

3-2.5 Piping, valves and fittings may have combustible or noncombustible linings.

> **Corrosion is a problem in piping, valves, and fittings. Plastic can be useful in preventing corrosion. This *Code* section allows plastic or other such materials to be used internally, as linings in piping, valves, or fittings.**

3-3 Pipe Joints.

3-3.1 Joints shall be made liquidtight and shall be either welded, flanged or threaded, except that listed flexible connectors may be used when installed in accordance with 3-3.2. Threaded joints shall be made up tight with a suitable thread sealant or lubricant. Joints in piping systems handling Class I liquids shall be welded when located in concealed spaces within buildings.

> **To prevent the loss of liquid, *all* joints shall be liquidtight. This is particularly critical with the more hazardous Class I flammable liquids to ensure against leakage. This explains the requirement for welded joints in concealed areas of a building.**

3-3.2 Pipe joints dependent upon the friction characteristics or resiliency of combustible materials for mechanical continuity or liquidtightness of piping shall not be used inside buildings. They may be used outside of buildings above or below ground. If used aboveground outside of buildings, the piping shall either be secured to prevent disengagement at the fitting, or the piping system shall be so designed that any spill resulting from disengagement could not unduly expose persons, important buildings or structures, and could be readily controlled by remote valves.

---

*Available from the American Society of Mechanical Engineers, United Engineering Center, 345 East 47th St., New York, NY 10017.

The additional restrictions on pipe joints which are dependent on friction or resiliency characteristics of combustible materials to maintain liquidtightness are included because of the susceptibility of these materials to soften under fire exposure conditions, and because frost heaves may move the piping and thus disengage the joint. This latter problem has been particularly noticeable in the northern areas of the United States, often in service stations. The *Code* does limit use of this type of pipe joint to the outside of buildings. If used aboveground, piping must be secured or specially designed to prevent exposure to humans, important buildings, or structures, in the event of failure. In addition, the piping system must be readily controlled by a remote valve(s). Flexible connectors may also be used under the same guidelines. Flexible connectors are often used in underground piping systems to avoid damage if the tank should move. Swing joints made up with screwed elbows have been and still are used on some underground tanks. Flexible connectors are sometimes used at service stations, between fixed piping systems and the gasoline dispenser.

3-4 Supports.

3-4.1 Piping systems shall be substantially supported and protected against physical damage and excessive stresses arising from settlement, vibration, expansion or contraction. The installation of nonmetallic piping shall be in accordance with the manufacturer's instructions.

Expansion rods or flexible connectors can be used to prevent undue stresses in piping systems. Plastic reinforced with glass fibers is one example of nonmetallic piping.

3-5 Protection Against Corrosion.

3-5.1 All piping for liquids, both aboveground and underground, where subject to external corrosion, shall be painted or otherwise protected. (*See 2-3.3 for protection of piping connected to underground tanks.*)

3-6 Valves.

3-6.1 Piping systems shall contain a sufficient number of valves to operate the system properly and to protect the plant. Piping systems in connection with pumps shall contain a sufficient number of valves to control properly the flow of liquid in normal operation and in the event of physical damage. Each connection to piping by which equipment such as tank cars, tank vehicles or marine vessels discharge liquids into storage tanks shall be provided with a check valve for automatic protection against back-flow if the piping arrangement is such that back-flow from the system is possible. (*See also 2-2.7.1.*)

The *Code* requirements for number of valves is quite broad and nonspecific. This is because of the variety of piping systems and the varying situations dictated by the occupancy and use. Each piping system must be evaluated on its own merits, based on good engineering practice. If consideration is given to general safety guidelines, and to the potential results of failure, a determination as to the number and placement of valves can be made. Check valves are required only if back-flow is possible.

3-7 Testing.

3-7.1 Unless tested in accordance with the applicable sections of ANSI B31, *American National Standard Code for Pressure Piping*,* all piping before being covered, enclosed or placed in use shall be hydrostatically tested to 150 percent of the maximum anticipated pressure of the system, or pneumatically tested to 110 percent of the maximum anticipated pressure of the system, but not less than 5 lbs per sq in. (34.48 kPa) gage at the highest point of the system. This test shall be maintained for a sufficient time to complete visual inspection of all joints and connections, but for at least 10 minutes.

Tests are not a guarantee that the system is perfect or that it will remain in good condition. They are, however, one method to help ensure that the system will perform as desired under normal circumstances and predictable emergency situations. Three methods are acceptable for testing: (1) meeting the requirements of ANSI B31,[1] (2) hydrostatic testing to 150 percent of maximum system design pressure, or (3) pneumatically testing to 110 percent of the maximum system design pressure. In any case, the test must be conducted for at least ten minutes or until a complete visual inspection is conducted of all joints and connections. The basic reason for the percentage differences between the pneumatic and hydrostatic test limits is safety. In the event of failure, a hydrostatic test will quickly relieve its pressure because of the noncompressibility of most fluids, and the escaping fluid (normally water) will safely discharge. A pneumatic failure will sometimes involve violent explosion of the system being tested.

REFERENCES CITED BY *CODE*

(*These publications comprise a part of the requirements to the extent called for by the* Code.)

ANSI B31.3, *Petroleum Refinery Piping*, American National Standards Institute, New York, 1976.

ANSI B31.4, *Liquid Petroleum Transportation Piping Systems*, American National Standards Institute, New York, 1974.

ASTM A-395-77, *Ferritic Ductile Iron Pressure Retaining Castings for Use at Elevated Temperatures*, American Society for Testing and Materials, Philadelphia, 1977.

REFERENCES CITED IN COMMENTARY

[1]**ANSI B31**, *American National Standard Code for Pressure Piping*, American National Standards Institute, New York.

---

*Available from the American Society of Mechanical Engineers, United Engineering Center, 345 East 47th St., New York, NY 10017.

4

Container and Portable Tank Storage

Chapter 4 is the last chapter that is general in scope. All subsequent chapters deal with specific types of occupancies.

Chapter 4 is based on studies, research, and fire tests which have determined the performance of storage in containers and portable tanks under various fire conditions. Appendices C and D explain and amplify some of the Chapter 4 requirements and also describe some of the relevant accidental fires and fire tests. Chapter 4 also contains information on: the design, construction, and capacity of containers in storage cabinets; the design, construction, and operation of inside storage areas; and fire protection and fire control requirements for both indoor and outdoor storages of flammable and combustible liquids in containers and portable tanks.

4-1 Scope.

4-1.1 This chapter shall apply to the storage of liquids, including flammable aerosols, in drums or other containers not exceeding 60 gal (227.1 L) individual capacity and portable tanks not exceeding 660 gal (2,498.1 L) individual capacity and limited transfers incidental thereto. For portable tanks exceeding 660 gal (2,498.1 L), Chapter 2 shall apply.

Provisions of Chapter 4 of this *Code* apply to the storage of liquids in vessels with two distinct size limitations: (1) containers up to 60 gal (227.1 L), and (2) portable tanks of 60 to 660 gal (227.1 to 2,498.1 L) capacity. This fact should be kept in mind when reading or applying sections of this chapter.

The chemical and distillery industries have been using tanks with capacities up to 5,500 gal (20,818 L) for shipping products. The tanks are built in accordance with U. S. Department of Transportation specifications, and are considered by the user industries to be portable tanks. Since the tanks do not meet the *Code* definition of portable tanks because of excessive quantity, the reader is referred to Chapter 2, Tank Storage, for guidance. As noted in Chapter 9 of the *Code*, refineries, chemical plants, and distilleries are exempt from the application of this chapter (Chapter 4).

4-1.2 This chapter shall not apply to the following:

(a) Storage of containers in bulk plants, service stations, refineries, chemical plants and distilleries.

The storage of containers in bulk plants, service stations, refineries, chemical plants, and distilleries is excluded because each such occupancy has its own storage and handling requirements. Requirements for the storage of containers in these particular occupancies are contained in Chapters 5 through 9 of this *Code*. Often, their storage requirements are complex because their operations inherently involve flammable and combustible liquids. Thus, their handling operations become as important as their storage considerations.

(b) Liquids in the fuel tanks of motor vehicles, aircraft, boats or portable or stationary engines.

(c) Beverages when packaged in individual containers not exceeding a capacity of one gallon.

(d) Medicines, foodstuffs, cosmetics and other consumer products containing not more than 50 percent by volume of water miscible liquids and with the remainder of the solution not being flammable when packaged in individual containers not exceeding one gallon in size.

(e) The storage of liquids that have no fire point when tested by ASTM D 92-72, the Cleveland Open Cup Test Method, up to the boiling point of the liquid, or up to a temperature at which the sample being tested shows an obvious physical change.

This may include physical changes such as the change from a liquid to a gas or a liquid to a solid, a change in color, or an obvious change in viscosity. The reader is again reminded that the fire point is the temperature at which sustained burning will take place upon application of an outside ignition source to the vapors. The fire point is usually several degrees higher than the flash point, but still several hundred degrees lower than the auto ignition temperature of the liquid. The liquids that are intended for exclusion by this section of the *Code* are those that show a flash point when tested in accordance with a Tag or other closed-cup test method, but which do not burn. Water-based paints, which are formulated with small amounts of flammable or combustible solvents, are one example. Although these products may exhibit a flash point, the end product will not support combustion until the water boils off and the residue becomes thick and highly viscous.

(f) The storage of distilled spirits and wines in wooden barrels or casks.

4-1.3 For the purpose of this chapter, unstable liquids and flammable aerosols shall be treated as Class IA liquids.

Even though some of these liquids would normally not be classified as Class

IA flammable liquids, additional requirements for their safe storage are recommended. Unstable liquids may decompose or react vigorously during a fire, causing the container to overpressurize and rupture much more rapidly than with stable liquids. Similarly, flammable aerosols are liquids in pressurized containers that are likely to rupture under fire exposure conditions, spreading fire as they rocket. See Section 1-2 for the definition of flammable aerosol.

4-2 Design, Construction, and Capacity of Containers.

4-2.1 Only approved containers and portable tanks shall be used. Metal containers and portable tanks meeting the requirements of, and containing products authorized by, Chapter I, Title 49 of the *Code of Federal Regulations* (DOT Regulations), or NFPA 386, *Standard for Portable Shipping Tanks*, shall be acceptable. Polyethylene containers meeting the requirements of, and containing products authorized by, DOT Specification 34, and polyethylene drums authorized by DOT Exemption Procedures, shall be acceptable. Plastic containers meeting the requirements of ANSI/ASTM D3435-78, *Plastic Containers (Jerry Cans) for Petroleum Products*, used for petroleum products within the scope of that specification shall be acceptable.

As with the larger storage tanks, the *Code* attempts to reduce fire hazards by making rigid requirements for the design and construction of containers and portable tanks. Thus, only approved containers and portable tanks shall be used. This *Code* section specifies what might be considered an approved container or portable tank. It is important to note and emphasize that NFPA does not approve any equipment, plans, or specifications. The authority having jurisdiction must specify such approval conditions.

4-2.2 Each portable tank shall be provided with one or more devices installed in the top with sufficient emergency venting capacity to limit internal pressure under fire exposure conditions to 10 psig (68.95 kPa), or 30 percent of the bursting pressure of the tank, whichever is greater. The total venting capacity shall be not less than that specified in 2-2.5.4 or 2-2.5.6. At least one pressure-actuated vent having a minimum capacity of 6,000 cu ft (169.92 m^3) of free air per hour [14.7 psia (101.3 kPa) and 60°F (15.6°C)] shall be used. It shall be set to open at not less than 5 psig (34.48 kPa). If fusible vents are used, they shall be actuated by elements that operate at a temperature not exceeding 300°F (148.9°C). When used for paints, drying oils and similar materials where plugging of the pressure-actuated vent can occur, fusible vents or vents of the type that soften to failure at a maximum of 300°F (148.9°C) under fire exposure may be used for the entire emergency venting requirement.

The portable tanks described in this paragraph are generally rectangular in shape, approximately 440 gal (1,676 L) capacity, and are used for transporting flammable paints and other finishing products. They may be constructed of steel, aluminum, or magnesium.

Vents are to be installed in the top of the tank in an effort to prevent the discharge of liquid under overpressure situations. Normally, vents are used for expelling vapors which are generated by increase in temperature, to avoid overpressurizing of the tank. There are two general types of venting requirements: (1) vents that are designed to minimize or eliminate overpressurizing due to normal changes in the ambient temperature, and (2) the emergency relief venting required when the tank is subjected to extreme temperature increases such as those generated by a fire situation. The requirements of this paragraph for 6,000 cu ft (169.92 m^3) of free air per hour may not be adequate for unstable liquids. It is assumed that tanks equipped with fusible vents will not exceed 5 psig by the time the container walls are heated to a temperature of 300 °F (148.9 °C).

4-2.3 Containers and portable tanks for liquids shall conform to Table 4-2.3 except as provided in 4-2.3.1 or 4-2.3.2.

Table 4-2.3
Maximum Allowable Size of Containers and Portable Tanks

Container Type	Flammable Liquids Class IA	Class IB	Class IC	Combustible Liquids Class II	Class III
Glass	1 pt	1 qt	1 gal	1 gal	5 gal
Metal (other than DOT drums) or approved plastic	1 gal	5 gal	5 gal	5 gal	5 gal
Safety Cans	2 gal	5 gal	5 gal	5 gal	5 gal
Metal Drum (DOT Spec.)	60 gal	60 gal	60 gal	60 gal	60 gal
Approved Portable Tanks	660 gal	660 gal	660 gal	660 gal	660 gal
Polyethylene DOT Spec. 34, or as authorized by DOT Exemption	1 gal	5 gal	5 gal	60 gal	60 gal

SI Units: 1 pt = 0.473 L; 1 qt = 0.95 L; 1 gal = 3.785 L.

An example of an acceptable high-density polyethylene plastic container is the "jerry" can which is designed to meet the ANSI/ASTM D3435-78[1] requirement, and which has been approved by some jurisdictions for storing petroleum products. Plastic containers made of low-density polyethylene, such as found in bleach-bottle type construction, do not meet the rigid requirements of the special high-density plastic containers. They should not be allowed for use with petroleum products.

4-2.3.1 Medicines, beverages, foodstuffs, cosmetics and other common consumer products, when packaged according to commonly accepted practices for retail sales, shall be exempt from the requirements of 4-2.1 and 4-2.3.

4-2.3.2 Class IA and Class IB liquids may be stored in glass containers of not more than one gallon capacity if the required liquid purity (such as ACS analytical reagent grade or higher) would be affected by storage in metal containers or if the liquid would cause excessive corrosion of the metal container.

This exception is included to avoid unnecessarily restricting research operations involving flammable and combustible liquids, while still providing an acceptable degree of fire safety. The keys in applying this *Code* section are: (1) that the liquid is to be stored, (2) that the purity of the liquid would be affected if stored in approved metal containers, and (3) to consider the effect of the liquid on the structural capabilities of the container.

4-3 Design, Construction and Capacity of Storage Cabinets.

4-3.1 Not more than 120 gal (454.2 L) of Class I, Class II and Class IIIA liquids may be stored in a storage cabinet. Of this total, not more than 60 gal (227.1 L) may be of Class I and Class II liquids and not more than three (3) such cabinets may be located in a single fire area, except that, in an industrial occupancy, additional cabinets may be located in the same fire area if the additional cabinet, or group of not more than three (3) cabinets, is separated from other cabinets or group of cabinets by at least 100 ft (30.48 m).

Depending upon the particular occupancy of a building, certain lesser quantities of flammable or combustible liquids are permitted to be stored in a safe place, outside of a specific storage cabinet or room. (See Section 4-5 and Chapters 5 through 7 for additional details as to when storage cabinets are required.) Large quantities will often be required to be stored in a separate storage room or area, or possibly may require storage in a flammable liquid warehouse.

Provisions covering locations where flammable or combustible liquids may be stored are arranged in the *Code* according to increasing size, from storage cabinets, inside rooms, cutoff rooms, and attached buildings, to warehouses.

Most commercially available and approved storage cabinets are built to hold 60 gal (227.1 L) or less of flammable and combustible liquids. A fire area is an area of a building separated from the rest of the building by construction having a fire resistance rating of at least 1 hr and having all communicating openings properly protected by an assembly having a fire resistance rating of at least 1 hr. Note that in industrial occupancies, the concept of fire area can be achieved by distance [100 ft (30.48 m)] as well as by construction.

Industrial occupancies are intended to include factories making products of all kinds, and properties devoted to operations such as processing, assembling, mixing, packaging, finishing or decorating, and repairing. (Also see 5-1.1)

4-3.2 Storage cabinets shall be designed and constructed to limit the internal temperature at the center, 1 in. (25.40 mm) from the top to not more than 325 °F (162.8 °C) when subjected to a 10-minute fire test with burners simulating a room fire exposure using the standard time-temperature curve as given in ASTM E152—72. All joints and seams shall remain tight and the door shall remain securely closed during the fire test. Cabinets shall be labeled in conspicuous lettering, "FLAMMABLE — KEEP FIRE AWAY."

> An approved storage cabinet is designed and constructed to protect the contents from external fires. The *Code* does not require that storage cabinets be vented. Some jurisdictions mandate that the interior of the storage cabinets be vented to minimize the accumulation of vapors inside the cabinet. However, venting a storage cabinet may defeat the cabinet's designed purpose of protecting any container of flammable or combustible liquids from involvement in a standard room fire for up to 10 minutes. That is the estimated time for a particular room or area of the building to be seriously involved in fire. When storage cabinets are provided with vents by manufacturers, the vents are plugged in order to pass the indicated fire tests.
>
> The referenced time-temperature curve as given in ASTM E-152-78[2] is also included in NFPA 252.[3] The points on that curve that determine its character are: 1,000 °F (538 °C) at 5 minutes; 1,300 °F (704 °C) at 10 minutes; 1,550 °F (843 °C) at 30 minutes; 1,638 °F (892 °C) at 45 minutes; 1,700 °F (927 °C) at 1 hour; 1,792 °F (978 °C) at 1½ hours; 1,925 °F (1,052 °C) at 3 hours. After 2 hours, the rate of rise is 75 °F (41.7 °C) per hour.

4-3.2.1 Metal cabinets constructed in the following manner are acceptable. The bottom, top, door and sides of cabinet shall be at least No. 18 gage sheet steel and double walled with 1½ in. (38.1 mm) air space. Joints shall be riveted, welded or made tight by some equally effective means. The door shall be provided with a three-point latch arrangement and the door sill shall be raised at least 2 in. (50.8 mm) above the bottom of the cabinet to retain spilled liquid within the cabinet.

> See Figure 4-1.
>
> The three-point latch arrangement on the doors is intended to help maintain the integrity of the cabinet under a fire exposure. Without a three-point latch arrangement, the metal will warp when exposed to a fire and expose the cabinet's contents to the fire. Note that the *Code* does not specify a test for cabinets built in accordance with this paragraph.

4-3.2.2 Wooden cabinets constructed in the following manner are acceptable. The bottom, sides and top shall be constructed of exterior grade plywood at least 1 in. (25.40 mm) in thickness, which shall not break down or delaminate under fire conditions. All joints shall be rabbetted and shall be fastened in two directions with wood screws. When more than one door is used, there shall be a rabbetted overlap of not less than 1 in. (25.40 mm). Doors shall be equipped with a means of latching and hinges shall be constructed and mounted in such a manner as to not

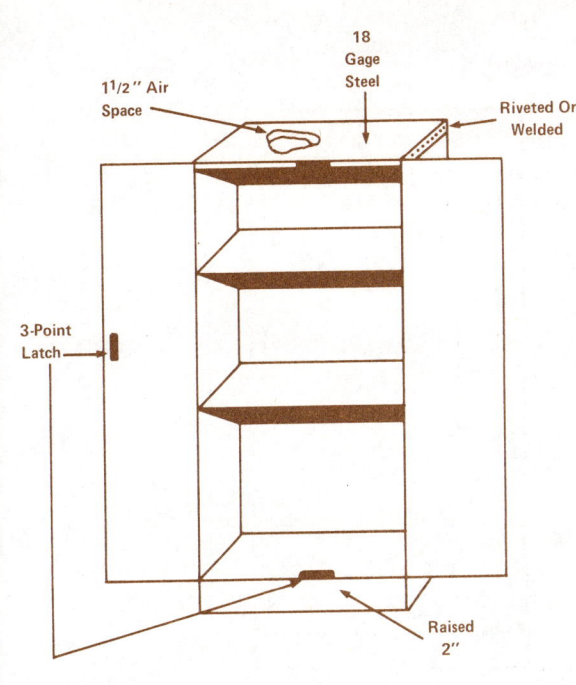

Figure 4-1. Metal cabinets constructed in this manner are acceptable.

lose their holding capacity when subjected to fire exposure. A raised sill or pan capable of containing a 2-in. (50.8-mm) depth of liquid shall be provided at the bottom of the cabinet to retain spilled liquid within the cabinet.

See Figure 4-2.

Although not required to do so, a cabinet built to these requirements will meet the test specifications of 4-3.2. Those built in accordance with 4-3.2.1 may not do so.

4-3.2.3 Listed cabinets which have been constructed and tested in accordance with 4-3.2 shall be acceptable.

4-4 Design, Construction and Operation of Separate Inside Storage Areas. (*See 1-2 Definitions.) (For additional information, see Appendix C.)*

The reader is referred to the definitions in Section 1-2 of "inside room," "cutoff room," and "attached building." Generally, inside rooms provide relatively large storage capacity for flammable and combustible liquids. In many occupancies, it is common for inside rooms to serve as intermediate storage locations. For example, many hospitals and large university science departments have an outside building that is used for storing their bulk quantities of flammable and combustible liquids. These outside storage buildings then feed several inside flammable and combustible liquid storage rooms which

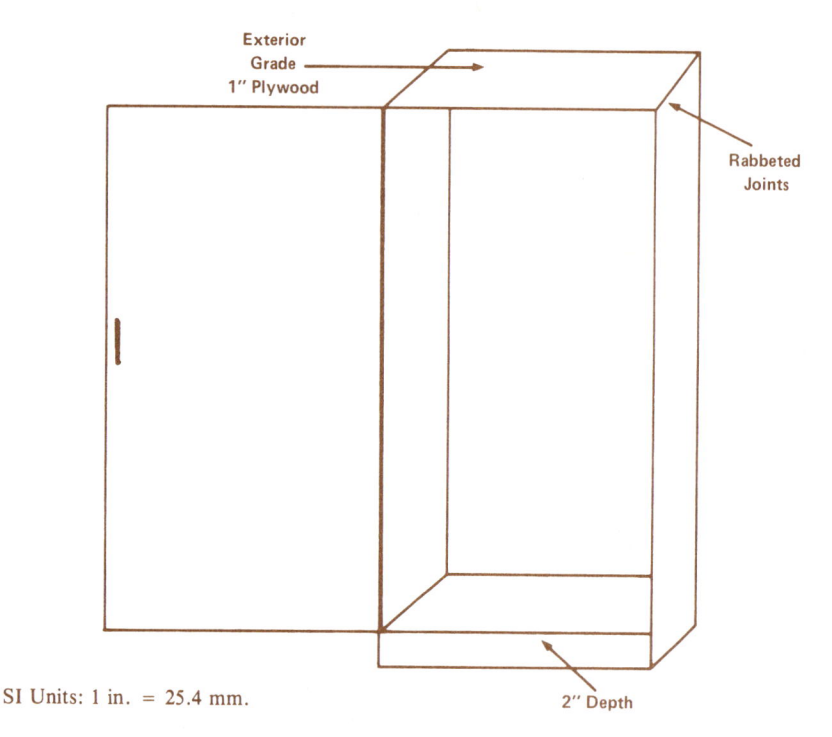

Figure 4-2. *Wooden cabinets constructed in this manner are acceptable. Note that the* Code *does not require a three-point latch on doors of wooden cabinets. When exposed to fire, wood does not tend to warp or distort.*

may be located on different floors of the main building. In turn, these inside storage rooms feed the storage cabinets in the individual laboratories where the liquids are finally used. Because of the potential fire hazard from inside storage areas to the rest of the building, the *Code* makes rigid requirements for the design, construction, and operation of these rooms. Note that explosion relief is not required for inside rooms, if all requirements are rigidly adhered to, even though Class I liquids may be dispensed within them.

4-4.1 Inside Rooms.

4-4.1.1 Inside rooms shall be constructed to meet the selected fire resistance rating as specified in 4-4.1.4. Such construction shall comply with the test specifications given in NFPA 251, *Standard Methods of Fire Tests of Building Construction and Materials.* Except for drains, floors shall be liquidtight and the room shall be liquidtight where the walls join the floor. Where an automatic fire protection system is provided, as indicated in 4-4.1.4, the system shall be designed and installed in accordance with the appropriate NFPA standard for the type of system selected.

4-4.1.2 Openings in interior walls to adjacent rooms or buildings shall be provided with:

(a) Normally closed, listed 1½ hr (B) fire doors for interior walls with fire resistance rating of 2 hr or less. Where interior walls are required to have greater than 2 hr fire resistance rating, the listed fire doors shall be compatible with the wall rating. Doors may be arranged to stay open during material handling operations if doors are designed to close automatically in a fire emergency by provision of listed closure devices. Fire doors shall be installed in accordance with NFPA 80, *Standard for Fire Doors and Windows*.

(B) rated doors are used in 2-hr fire-rated partitions which are to provide horizontal fire separation. The time rating indicates the duration of the test exposure. Note that the test exposure is determined according to the procedure in NFPA 251[4] and ASTM E-119-79[5].

(b) Noncombustible liquidtight raised sills or ramps at least 4 in. (101.6 mm) in height or otherwise designed to prevent the flow of liquids to the adjoining areas. A permissible alternative to the sill or ramp is an open-grated trench, which drains to a safe location, across the width of the opening inside of room.

Ramping is the most common method. If an open-grated trench is used, it must extend across the entire opening (door) into the room and must be located on the inside of the room. The trench must also be designed to drain the collected spilled liquid to a safe location. This method may be desirable if there is an extensive need to transfer flammable liquids in and out of the room in large quantities by means of hand trucks. (See Figure 4-3.)

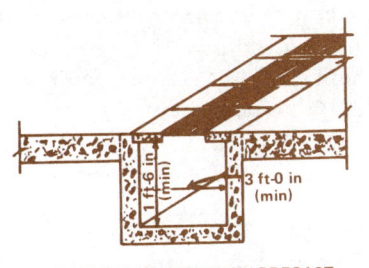

DRAINAGE TRENCH WITH PRECAST CONCRETE AND STEEL GRATING COVER

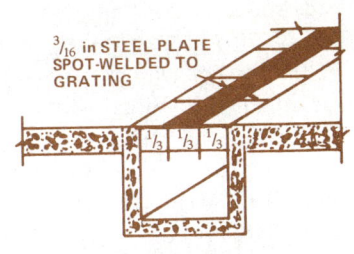

DRAINAGE TRENCH WITH STEEL PLATE AND GRATING COVER

SI Units: 1 in. = 25.4 mm; 1 ft = 0.3048 m.

Figure 4-3. Drainage trench details.

4-4.1.3 Wood at least 1 in. (25.40 mm) nominal thickness may be used for shelving, racks, dunnage, scuffboards, floor overlay and similar installations.

Note the permissive language of the *Code* in that wood may be used in an inside room. While the wood adds to the fire loading of the room, it is often desirable to use wood for shelving within the room. The *Code* does not mention the use of metal for such installations. There is no control over its use. Wood is preferable because it minimizes the chances of mechanical damage and sparks. If the room has been properly designed, the ventilating system should normally

maintain the atmosphere below the lower flammable limits of the stored materials. However, since we are dealing with flammable and combustible liquids, there is always the possibility of fire. Therefore, all precautions must be taken to prevent the ignition of the stored product.

4-4.1.4 Storage in inside rooms shall comply with the following:

Automatic Fire Protection* Provided	Fire Resistance	Maximum Floor Area	Total Allowable Quantities — Gallons/ Sq Ft/Floor Area
YES	2 hr	500 sq ft	10
NO	2 hr	500 sq ft	4**
YES	1 hr	150 sq ft	5
NO	1 hr	150 sq ft	2

SI Units: 1 sq ft = 0.0929 m².

*Fire protection system shall be sprinkler, water spray, carbon dioxide, dry chemical, halon or other approved system.

**Total allowable quantities of Class 1A and 1B Liquids shall not exceed that permitted in Table 4-4.2.7 and the provisions of 4-4.2.9.

The second footnote (**) was added to the 1981 Edition of the *Code*. An inside room without automatic fire protection could have had as much as 2,000 gal (7570 L) of a Class IA or Class IB liquid, while the maximum quantity is limited to 660 gal (2498 L) and 1,375 gal (5204 L), respectively in a cutoff room or attached building. This change reconciles that discrepancy.

Note that if automatic fire protection is not provided, the allowable quantities of flammable and combustible liquids to be stored in the room are severely curtailed. With construction of reduced fire resistance, the total allowable quantity stored is reduced from 5,000 gal (18 925 L) to 2,000 gal (7570 L), or from 750 gal (2839 L) to 300 gal (1136 L), depending on the fire resistance rating; if construction is reduced in fire resistance rating, the total allowable quantity is reduced from 5,000 gal (18 925 L) to 750 gal (2839 L), even if automatic fire protection is provided.

4-4.1.5 Electrical wiring and equipment located in inside rooms used for Class I liquids shall be suitable for Class I, Division 2 classified locations; for Class II and Class III liquids, shall be suitable for general use. NFPA 70, the *National Electrical Code*®, provides information on the design and installation of electrical equipment.

Caution: All "vaporproof" types of lighting fixtures are not suitable for use in atmospheres containing flammable or combustible liquid vapors. This designation often refers only to the fixture's capability to prevent moisture from entering the fixture. Approved fixtures will be properly labeled for their use and will normally be indicated as being "vaportight" or "explosionproof."

4-4.1.6 Every inside room shall be provided with either a gravity or a continuous mechanical exhaust ventilation system. Mechanical ventilation shall be used if Class I liquids are dispensed within the room.

(a) Exhaust air shall be taken from a point near a wall on one side of the room and within 12 in. (304.8 mm) of the floor with one or more make-up inlets located on the opposite side of the room within 12 in. (304.8 mm) from the floor. The location of both the exhaust and inlet air openings shall be arranged to provide, as far as practicable, air movements across all portions of the floor to prevent accumulation of flammable vapors. Exhaust from the room shall be directly to the exterior of the building. If ducts are used, they shall not be used for any other purpose and shall comply with NFPA 91, *Standard for the Installation of Blower and Exhaust Systems for Dust, Stock and Vapor Removal or Conveying*. If make-up air to a mechanical system is taken from within the building, the opening shall be equipped with a fire door or damper, as required in NFPA 91, the *Standard for the Installation of Blower and Exhaust Systems for Dust, Stock, and Vapor Removal or Conveying*. For gravity systems, the make-up air shall be supplied from outside the building.

Ventilation is vital to the prevention of flammable liquid fires and explosions. In a room where the potential accumulation of flammable vapors can fall within the flammable limits of the stored material, every effort must be made to confine, remove, or dilute these vapors, thereby reducing the probability of a fire occurrence. Remember that the vapors from most flammable and combustible liquids are heavier than air (the only exception, in NFPA 325M,[6] is for HCN), thus the requirement for making sure that exhaust ventilation is initiated at, or near, the floor level.

(b) Mechanical ventilation systems shall provide at least one cubic foot per minute (4.25 m^3) of exhaust per square foot of floor area, but not less than 150 CFM. The mechanical ventilation system for dispensing areas shall be equipped with an airflow switch or other equally reliable methods which is interlocked to sound an audible alarm upon failure of the ventilation system.

When relying on a mechanical ventilation system, it is important to know that the system is functioning at all times. This explains the requirement for the air-flow (or other) interlocking switch, since air movement less than the intended design might allow the accumulation of a hazardous level of flammable vapors in the room. Also see comments following 5-3.5.1.

4-4.1.7 In every inside room, an aisle at least 3 ft (0.9144 m) wide shall be maintained so that no container is more than 12 ft (3.66 m) from the aisle. Containers over 30 gal (113.5 L) capacity storing Class I or Class II liquids shall not be stored more than one container high.

> Easy movement within the room is necessary in order to reduce the potential for spilling or damaging the containers and to provide both access for fire fighting and ready escape paths for occupants of the room should a fire occur. To further reduce the possibility of damage to containers, Class I and II flammable liquid containers over 30 gal (113.5 L) in capacity should not be stored more than one container high. Such containers are built to DOT specifications and are not required to withstand a drop test greater than 3 ft (0.9144 m) when full.

4-4.1.8 Where dispensing is being done in inside rooms, operations shall comply with the provisions of Chapter 5.

4-4.1.9 Basement Storage Areas. Class I liquids shall not be permitted in inside storage rooms in basement areas.

> This is a new section of the *Code*. Since 4-4.2.11 prohibits Class I storage in basements, a comparable restriction should apply to inside rooms wherever practical.

4-4.2 Cutoff Rooms and Attached Buildings.

4-4.2.1 Construction design of exterior walls shall provide ready accessibility for fire fighting operations through provision of access openings, windows or lightweight noncombustible wall panels. Where Class IA or IB liquids are dispensed, or where Class IA liquids are stored in containers larger than one gallon, the exterior wall or roof construction shall be designed to include explosion venting features, such as lightweight wall assemblies, light-weight roof assemblies, roof hatches or windows of the explosion venting type. NFPA 68, *Guide for Explosion Venting*, provides information on this subject.

> Because some operations require storage of a greater quantity of flammable and combustible liquids than would be allowed in an inside storage room, a cutoff room or attached building may be used. These latter areas allow a greater quantity of liquids to be stored because they have at least one exterior wall. The exterior wall construction affords ready accessibility for fire fighting operations and explosion venting of the room. Explosion venting is required because of the increased hazard presented by the dispensing of Class IA and IB liquids, and the volatility of Class IA liquids. This technique is designed to limit pressures developed by the explosive ignition of flammable vapors and to prevent or reduce structured or mechanical damage. Consideration must be given to the direction in which an explosion is to be vented in order to prevent injury to personnel and damage to exposures.

CONTAINER AND PORTABLE TANK STORAGE 113

4-4.2.2 Where other portions of buildings or other properties are exposed, each opening in the exposing wall shall be protected with a listed 1½ hr (D) fire door installed in accordance with NFPA 80, *Standard for Fire Doors and Windows*, and the walls shall have a fire resistance rating of not less than 2 hr.

The (D) rating on the door indicates that the door is designed for installation in an exterior wall which is subject to severe fire exposure. The openings for explosion venting should not be located in the same wall that requires the rated door.

4-4.2.3 Except as noted in 4-4.2.6, interior walls, ceiling and floors shall have a fire resistance rating of not less than 2 hr where floor area of the room or building exceeds 300 sq ft (27.87 m²) or a fire resistance rating of not less than one hour for a floor area of 300 sq ft (27.87 m²) or less. Such construction shall comply with the test specifications given in NFPA 251, *Standard Methods of Fire Tests of Building Construction and Materials*. Walls shall be liquidtight at the floor level.

Note that the exception in 4-4.2.6 applies only to roofs of attached buildings, not to cutoff rooms.

4-4.2.4 Openings in interior walls to adjacent rooms or buildings shall be in accordance with 4-4.1.2 (a).

Any opening in the interior wall of cutoff rooms or attached buildings to adjacent rooms or buildings must have fire doors and either a drainage trench or raised sills or ramps that meet the same requirements as those for inside rooms. See comments under 4-4.1.2.

4-4.2.5 Curbs, scuppers, special drains or other suitable means shall be provided to prevent the flow of liquids under emergency conditions into adjacent building areas except where the individual container capacity is 5 gal (18.9 L) or less or if the liquids stored are only Class III liquids. The drainage system, if used, shall have sufficient capacity to carry off expected discharge of water from fire protection systems and hose streams.

The exceptions reflect the reduced hazard involved with smaller quantities of flammable and combustible liquids and higher flash point liquids. The reader is cautioned to make sure that the most restrictive case is considered in the initial construction of these rooms since later remodeling in order to upgrade the facility for use with more hazardous liquids can be expensive. The owner and users of cutoff rooms and attached buildings which lack curbs, scuppers, etc., must be advised so they understand that 5 gal (18.9 L) is the maximum allowable size of the individual container for any Class I or Class II flammable liquid. Because most flammable liquids immiscible with water are lighter than water, it is important that the capacity of the drainage system be sufficient to drain the water from the fire protection system and the hose streams, as well as the flammable liquid itself.

4-4.2.6 Roofs of attached buildings, one story in height, may be lightweight noncombustible construction if the separating interior wall as specified in 4-4.2.3 has a minimum 3 ft (0.9144 m) parapet.

4-4.2.7 Unprotected storage in cutoff rooms and attached buildings shall comply with Table 4-4.2.7. (*See 4-4.2.10 for mixed storage of liquids.*)

The referenced table is to be used with flammable and combustible liquids stored in cutoff rooms or attached buildings not equipped with an automatic fire protection system. This table specifies the maximum pile height, maximum quantity per pile, and the maximum total quantity allowed for each class of liquid. When applying the provisions of this table to actual situations, it is important to consider each restricting limitation in succession to avoid overlooking individual limitations. Therefore, the first item to consider would be the maximum allowable pile height in accordance with the class of the liquid, then the maximum quantity per pile, and lastly, the maximum total quantity within the cutoff room or attached building. Note that there are different requirements for container storage than for portable tank storage. Larger quantities are permitted in portable tanks because they are vented for fire exposure conditions and are less likely to explode from internal overpressure. Containers and portable tanks cannot be in the same pile because of this fact.

In order to put Table 4-4.2.7 to practical use, three examples follow:

1. You are in charge of the chemical supply room of a large state university. You have just received 900 gal (3406.5 L) of isopropyl alcohol (Class IB) and 900 gal (3406.5 L) of 95 percent ethyl alcohol (Class IB), all in containers. To try to conserve space, you arrange each liquid in two piles of 450 gal (1703.2 L) per pile.

Solution: Both alcohols are Class IB liquids, so the 450 gal (1703.2 L) per pile is permissible. However, Table 4-4.2.7 allows a maximum total quantity of 1,375 gal (5204 L). The received quantity of flammable liquids is 1,800 gal (6813 L), however, and this is beyond the amount permitted.

2. On an inspection of a wire coating and insulation factory, you note that, among the many liquids in an attached building, there are four piles of portable tanks of a Class II liquid. Each pile is 6 ft (1.8 m) high and contains three 500 gal (1892.5 L) portable tanks.

Solution: Both the quantity per pile [1,500 gal (5700 L)] and the total quantity [6,000 gal (22 710 L)] are under the maximum quantities permitted for Class II liquids.

CONTAINER AND PORTABLE TANK STORAGE

Table 4-4.2.7 Indoor Unprotected Storage of Liquids in Containers and Portable Tanks

Class	Container Storage			Portable Tank Storage		
	Max. Pile Height (ft.)	Max. Quant. per Pile (gal.)	Max. Total Quant. (gal.)	Max. Pile Height (ft.)	Max. Quant. per Pile (gal.)	Max. Total Quant. (gal.)
1A	5	660	660	—	Not Permitted	—
1B	5	1,375	1,375	7	2,000	2,000
1C	5	2,750	2,750	7	4,000	4,000
II	10	4,125	8,250	7	5,500	11,000
IIIA	15	13,750	27,500	7	22,000	44,000
IIIB	15	13,750	55,000	7	22,000	88,000

SI Units: 1 ft = 0.3048 m; 1 gal = 3.785 L.

3. You have been hired by an insurance company to inspect the liquid storage room of a large manufacturing plant. In the storage room are eight small piles [under 5 ft (1.5 m) high] of containers storing degreasers with flash points above 140°F (60°C). Each pile contains 1,100 gal (4163.5 L) of the liquid.

Solution: The total amount of degreaser stored is substantially less than the maximum of 13,750 gal (52 040 L) specified for a pile storage of a Class IIIA liquid in Table 4-4.2.7.

4-4.2.8 Protected storage in cutoff rooms and attached buildings shall comply with Section 4-6 as applicable. (*See 4-4.2.10 for mixed storage of liquids.*)

Note that this *Code* paragraph refers to Section 4-6 for protected storage in cutoff rooms and attached buildings. The storage arrangement must meet the requirements for palletized or solid pile storage of flammable and combustible liquids in containers and portable tanks, or the requirements for the rack storage of liquids in containers. As is to be expected, larger amounts can be stored where protection is provided.

4-4.2.9 Wood at least 1 in. (25.4 mm) nominal thickness may be used for shelving, racks, dunnage, scuffboards, floor overlay and similar installations.

This change was made so that wood would be acceptable in cutoff rooms and attached buildings, as it is in inside rooms. See 4-4.1.3.

4-4.2.10 Where two or more classes of liquids are stored in a single pile or rack section, the maximum quantities and height of storage permitted in that pile or rack section shall be the smallest of the two or more separate quantities and heights. The maximum total quantities permitted shall be limited to a sum of proportional amounts that each class of liquid present bears to the maximum total permitted for its respective class; sum of proportional amounts not to exceed 100 percent.

Because containers of volatile liquids are more likely to rupture under fire conditions, the *Code* restricts the maximum quantities and height of mixed storage in a pile or rack section to the more restrictive requirement. For example, when both Class I and Class II flammable liquid containers are stored in a single pile within an unprotected cutoff room, the maximum allowable pile height would be 5 ft (1.5 m).

When calculating the maximum total quantity of each particular class allowed, one should start with the most restrictive requirement, which is that of the lower class flammable liquid. Compute the percentage of this amount as compared to the maximum total quantity allowed for that particular class. Add the proportional, or percentage amounts, of each class. The sum total of these percentages (or proportional amounts) cannot exceed 100. For example, 1,000 gal (3785 L) of Class IB flammable liquids in portable tanks would represent 50 percent of the allowable quantities for Class IB liquids. One would then be allowed to add 50 percent of another class within this same cutoff room or attached building — such as 330 gal (1249 L) of Class IA flammable liquids in containers — or other combinations which would total 50 percent of the category's maximum allowable amount. The total of these percentages must not exceed 100. Application of this provision is illustrated in the following examples:

1. Calculating maximum total quantity in accordance with Table 4-4.2.7.

Containers of Class IA, Class IC, and Class IIIA liquids

Class IA	165 gal	$\frac{165}{660} = 25\%$
Class IC	550 gal	$\frac{550}{2,750} = 20\%$
Class IIIA	2,750 gal	$\frac{2,750}{27,500} = 10\%$

Portable tanks of Class IB and Class II liquids

Class IB	500 gal	$\frac{500}{2,000} = 25\%$
Class II	2,200 gal	$\frac{2,200}{11,000} = 20\%$
		TOTAL = 100%

2. What is the maximum quantity of a Class IIIA liquid that can be stored in portable tanks in an unprotected building that also contains 2,000 gal (7570 L) of a Class IC liquid in portable tanks?

Solution: According to Table 4-4.2.7, 2,000 gal (7570 L) of a Class IC liquid is 50 percent of the total of Class IC permitted [4,000 gal (15 140 L)] in portable tank storage. Since the sum of proportional amounts cannot exceed 100 percent, one can store 50 percent of the allowable amount of Class IIIA. This maximum amount is 44,000 gal (166 540 L), so 22,000 gal (83 270 L) of Class IIIA can be stored in the same unprotected building as 2,000 gal (7570 L) of a Class IC liquid.

3. What is the maximum amount of Class IA liquids that can be stored in an unprotected building containing 1,650 gal (6245.3 L) of a Class II liquid, assuming container storage?

Solution: 1,650 gal (6245.3 L) = 20 percent of 8,250 gal (31 226.3 L). This means that 80 percent of the allowable amount of Class IA can be stored. Eighty percent of 660 gal (2498 L) = 528 gal (1998.5 L) of Class IA liquid.

4-4.2.11 Dispensing operations of Class I or Class II liquids are not permitted in cutoff rooms or attached buildings exceeding 1000 sq ft (92.9 m^2) floor area. In rooms where dispensing of Class I liquids is permitted, electrical systems shall comply with 4-4.1.5, except that within 3 ft (0.914 m) of a dispensing nozzle area, the electrical system shall be suitable for Class I Division I; ventilation shall be provided per 4-4.1.6; and operations shall comply with the provisions of Chapter 5.

The rationale here is that dispensing introduces a hazard to the storage situation. Therefore, the size of the room where dispensing is allowed is restricted in size. Where dispensing takes place, other restrictions are imposed in addition to the limitation on room size.

4-4.2.12 Basement Storage Areas. Class I liquids shall not be permitted in the basement areas of cutoff rooms and attached buildings. Class II and Class IIIA liquids may be stored in basements provided that automatic sprinkler protection and other fire protection facilities are provided in accordance with Section 4-6.

Basement storage of any kind presents a problem in fire fighting. Access to most basements is restricted, and ventilation of the area can be more difficult to achieve. The storage itself adds to the fire loading and further restricts movement. If storage of Class I liquids were allowed, any vapors present would impose additional major difficulties. For all of these reasons, storage of Class I liquids is not allowed, and storage of combustible liquids requires that the basement be protected.

4-5 Indoor Storage.

4-5.1 Basic Conditions.

4-5.1.1 The storage of any liquids shall not physically obstruct a means of egress. Class I liquids in other than separate inside storage areas or warehouses shall be so placed that a fire in the liquid storage would not preclude egress from the area.

4-5.1.2 The storage of liquids in containers or portable tanks shall comply with 4-5.2 through 4-5.7 as applicable. Where separate inside storage areas are required, they shall conform to Section 4-4. Where other factors substantially increase or decrease the hazard, the authority having jurisdiction may modify the quantities specified.

4-5.1.3 Liquids used for building maintenance painting or other similar infrequent maintenance purposes may be stored temporarily in closed containers outside of storage cabinets or separate inside storage areas, if limited in amount, not to exceed a 10-day supply at anticipated rates of consumption.

These three paragraphs contain wording which is sometimes referred to as "code language," and it is worthy of comment. The words "so placed," "substantially," "infrequent," "temporarily," and "limited in amount" all imply that judgment is needed. Where such judgment is called for, the often-mentioned "authority having jurisdiction" enters the picture.

CONTAINER AND PORTABLE TANK STORAGE 119

Note the latitude given such an authority in 4-5.1.2, where he may make modifications based on factors that increase or decrease the hazard. This suggests that the modification may be *more or less* rigid than *Code* specifications, based on his assessment of the particular situation.

4-5.2 Dwellings and Residential Buildings Containing Not More Than Three Dwelling Units and Accompanying Attached and Detached Garages. Storage in excess of 25 gal (94.63 L) of Class I and Class II liquids combined shall be prohibited. In addition, storage in excess of 60 gal (227.1 L) of Class IIIA liquid shall be prohibited.

A proposal to reduce the allowable amount to 10 gal (37.85 L) of Class I and II liquids was considered by the Technical Committee. The conclusion was reached that 25 gal (94.6 L) is a reasonable amount for three dwelling units.

4-5.3 Assembly Occupancies, Buildings Containing More Than Three Dwelling Units, and Hotels. Storage in excess of 10 gal (37.85 L) of Class I and Class II liquids combined or 60 gal (227.1 L) of Class IIIA liquids shall be in containers stored in storage cabinets, in safety cans, or in a separate inside storage area not having an opening communicating with that portion of the building used by the public.

Because of the size of these buildings, there may be a need for larger quantities of flammable and combustible liquids. Given that increase, additional precautions are prescribed. The quantity restrictions are considered to be *total* quantities and not individual quantities for each dwelling unit or room to be occupied.

4-5.4 Office, Educational and Institutional Occupancies. Storage shall be limited to that required for operation of office equipment, maintenance, demonstration and laboratory work. This storage shall comply with the provisions of 4-5.4.1 through 4-5.4.4 except that the storage for industrial and educational laboratory work shall comply with NFPA 45, *Standard on Fire Protection for Laboratories Using Chemicals*.

This is a general requirement for total storage, requiring a judgmental assessment by the authority having jurisdiction. The next three paragraphs deal with specific storage requirements outside of storage cabinets or storage rooms.

4-5.4.1 Containers for Class I liquids outside of a separate inside storage area shall not exceed a capacity of 1 gal (3.785 L) except that safety cans can be of 2 gal (7.57 L) capacity.

Implicit in this requirement is the assessment that the safety can is the preferred type of storage container for these occupancies.

4-5.4.2 Not more than 10 gal (37.85 L) of Class I and Class II liquids combined shall be stored in a single fire area outside of a storage cabinet or a separate inside storage area unless in safety cans.

> This means, simply, that a maximum of ten 1 gal (3.785 L) containers is permitted outside of a storage cabinet or storage room per fire area. If the containers are safety cans, more than 10 gal (37.85 L) is allowed.

4-5.4.3 Not more than 25 gal (94.63 L) of Class I and Class II liquids combined shall be stored in a single fire area in safety cans outside of a separate inside storage area or storage cabinet.

> Even if the storage is in safety cans, 25 gal (94.63 L) is the maximum allowed per fire area outside of storage cabinets or a storage room.

4-5.4.4 Not more than 60 gal (227.1 L) of Class IIIA liquids shall be stored outside of a separate inside storage area or storage cabinet.

4-5.5 Mercantile Occupancies and Retail Stores and Other Related Areas Accessible to the Public.

4-5.5.1 In rooms or areas accessible to the public, storage of Class I, Class II and Class IIIA liquids shall be limited to quantities needed for display and normal merchandising purposes but shall not exceed 2 gal per sq ft (7.57 L per 0.0929 m^2) of gross floor area. Storage of Class IA liquids shall be prohibited in basement display areas and limited to 1 gal per sq ft (3.785 L per 0.0929 m^2) on other floors. In areas not protected, storage of Class IB, IC and II liquids on other than the ground floor shall be limited to 1 gal per sq ft (3.785 L per 0.0929 m^2) of gross floor area. Protected shall mean protected with automatic sprinklers installed at least in accordance with NFPA 13, *Standard for the Installation of Automatic Sprinklers*, requirements for ordinary hazard Group 2 occupancies. The gross floor area used for computing the maximum quantity permitted shall be considered as that portion of the floor actually being used for merchandising liquids and immediately adjacent aisles.

> The *Code* separates the requirements relating to the amount of flammable liquids needed for normal merchandising from the quantity needed for inventory. The inventory area is not normally accessible to the public. Note that the *Code* specifically defines what is meant by the terms "protected" and "gross floor area." In computing the gross floor area used for merchandising flammable and combustible liquids, only that portion of the shelving that is actually used for the merchandising of flammable liquids is to be considered. For example, in a department store that has 50 ft (15.2 m) of shelving on which all types of products are displayed and only 5 ft (1.5 m) of that shelving is used to display flammable and combustible liquids, only the 5 ft (1.5 m) length shall be used to

compute the gross floor area. If the shelving is 2 ft (0.6 m) deep and the aisle in front of the display is 4 ft (1.2 m) wide, the total width would be 6 ft (1.8 m). This would result in 30 ft (9 m) of gross floor area. If 2 gal/sq ft (7.6 L/0.09 m^2) is permitted, the *Code* would allow 60 gal/sq ft (227.1 L/0.09 m^2) of liquid to be stored in this area of shelving. In those areas where only 1 gal/sq ft (3.785 L/0.09 m^2) of gross floor area is allowed, the area would be limited to displaying a total of 30 gal (113.5 L). (See Figure 4-4.)

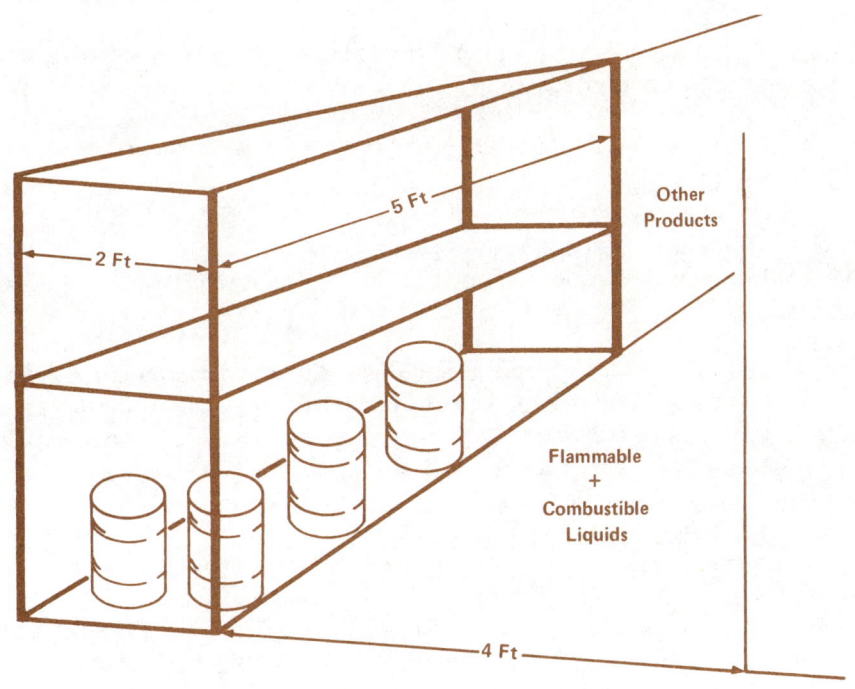

30 Ft^2 = GROSS FLOOR AREA

SI Units: 1 ft = 0.3048 m; 1 sq ft = 0.0929 m^2.

Figure 4-4. In computing the gross floor area used for merchandising flammable and combustible liquids, only that portion of the shelving that is actually used for the merchandising of flammable liquids is to be considered.

4-5.5.2 The aggregate quantity of additional stock in areas not accessible to the public shall not exceed the greater of that which would be permitted if the area were accessible to the public, or 60 gal (227.1 L) of Class IA, 120 gal (454.2 L) of Class IB, 180 gal (681.3 L) of Class IC, 240 gal (908.4 L) of Class II, or 660 gal (2498.1 L) of Class IIIA liquids, or 240 gal (908.4 L) in any combination of Class I and Class II liquids subject to the limitations of the individual class. These quantities may be doubled for areas protected as defined in 4-5.5.1. Storage of Class IA liquids shall be prohibited in basement storage areas.

This paragraph refers to the aggregate quantity of additional stock of flammable and combustible liquids used to replenish normal merchandising displays in areas not accessible to the public. This paragraph still applies the 2 gal/sq ft (7.6 L/0.09 m²) of gross floor area for protected buildings, and 1 gal/sq ft (3.785 L/0.09 m²) of floor area for unprotected buildings, as stated in 4-5.5.1. In addition, maximum quantity limitations are set forth by this paragraph. Note that these maximum quantity limitations can be doubled if the area is protected with an automatic sprinkler system installed in accordance with NFPA 13[7] requirements for ordinary hazard Group 2 occupancies.

4-5.5.3 Quantities in excess of those permitted in 4-5.5.2 shall be stored in accordance with other appropriate sections of this code.

Such storage would be in inside storage rooms, cutoff rooms, or attached buildings.

4-5.5.4 Containers shall not be stacked more than 3 ft (.9144 m) or 2 containers high, whichever is the greater, unless on fixed shelving or otherwise satisfactorily secured.

This paragraph limits display conditions so that the likelihood of containers falling or being knocked over is minimal. It is not the intent of this paragraph to allow unlimited height or stacking of containers on top of each other when on fixed shelving. The exception indicates that if the containers are placed on fixed shelving, the number of shelves is not limited. Containers 6 in. (152.4 mm) high can be stacked six high. Containers 15 in. (381 mm) high can only be stacked two high.

4-5.5.5 Shelving shall be of stable construction, of sufficient depth and arrangement such that containers displayed thereon shall not easily be displaced.

4-5.5.6 Leaking containers shall be removed immediately to an adequately ventilated area, and the contents transferred to an undamaged container.

4-5.6 General Purpose Warehouses. (*See 1-2 Definitions.*)

4-5.6.1 General purpose warehouses shall be separate, detached buildings or shall be separated from other type occupancies by a standard 4 hr fire wall, or, if approved, a fire partition having a fire resistance rating of not less than 2 hr. Each opening in a fire wall shall be protected with an automatic closing listed 3 hr (A) fire door with the fusible link or other automatic actuating mechanism located in the opening or on both sides of the opening. Each opening in a fire partition shall be protected with an automatic-closing listed 1½ hr (B) fire door. The doors shall be installed in accordance with NFPA 80, *Standard for Fire Doors and Windows.*

Some confusion has arisen in the past regarding the apparent equating of a fire wall and a fire partition. This was not the intent of the *Code*. The key word is "separation." If a 4-hr wall is needed, based on the circumstances, then only a fire wall shall be used. If a 2-hr fire partition will suffice, then the fire partition should be used. The words "if approved" have been inserted in the 1981 edition of the *Code* to try to clarify the paragraph and to avoid any misinterpretation.

The (A) rated fire doors are intended for use in openings in walls separating buildings or dividing a single building into fire areas. The 3-hr rating indicates a 3-hr test exposure. The (B) rated door is intended for use in a 2-hr fire rated partition providing horizontal fire separation. (See 4-4.1.2.)

Fire partitions do not necessarily extend from the basement through the roof. They may be used to divide a floor or area into smaller segments and extend only from the floor to the underside of the floor above. On the other hand, a fire wall is required to extend from the basement through all floors and the roof and thus *completely* separate two building areas, or two buildings.

4-5.6.2 Warehousing operations that involve storage of liquids shall be restricted to separate inside storage areas or to liquid warehouses in accordance with Section 4-4 or 4-5.7, as applicable, except as provided in 4-5.6.3.

4-5.6.3 Class IB and IC liquids in containers of 1 gal (3.785 L) or less capacity, Class II liquids in containers of 5 gal (18.93 L) or less capacity, Class III liquids in containers of 60 gal (227.1 L) or less capacity may be stored in warehouses handling combustible commodities, as defined in the scope of NFPA 231, *Standard for Indoor General Storage*, provided that the storage area is protected with automatic sprinklers in accordance with the provisions of this standard for 20 ft (6.1 m) storage of Class IV commodities and the quantities and height of liquid storage are limited to:

 (a) Class IB & IC 660 gal (2498.1 L) — 5 ft (1.5 m) high
 (b) Class II 1375 gal (5204 L) — 5 ft (1.5 m) high
 (c) Class IIIA 2750 gal (10409 L) — 10 ft (3.0 m) high
 (d) Class IIIB 13,750 gal (52044 L) — 15 ft (4.6 m) high

The liquid storage shall also conform to 4-5.6.4, 4-5.6.5, 4-5.6.6 and 4-5.6.7.

Note that Class IA liquids are not allowed to be stored in a general purpose warehouse. Some examples of Class IV commodities are: small appliances, typewriters, cameras with plastic parts, telephones, vinyl floor tiles, and other similar products that are packed and shipped in ordinary corrugated cartons that have protective filler materials made from plastic products.

4-5.6.4 Basement Storage Areas. Class I liquids shall not be permitted in the basement areas of buildings. Class II and Class IIIA liquids may be stored in basements provided that automatic sprinkler protection and other fire protection facilities are provided in accordance with Section 4-6.

See comment following 4-4.2.12.

4-5.6.5 Palletized, Solid Pile or Rack Storage. Liquids in containers may be stored on pallets, in solid piles or on racks subject to the quantities and heights limits of 4-5.6.3 provided the protection is in accordance with Section 4-6, as applicable.

4-5.6.6 Separation and Aisles. Palletized or solid pile storage shall be arranged so that piles permitted in 4-5.6.3 are separated from each other by at least 4 ft (1.2 m) aisles. Aisles shall be provided so that no container is more than 12 ft (3.6 m) from an aisle. Where the storage of liquids is on racks, a minimum 4 ft (1.2 m) wide aisle shall be provided between adjacent rows of racks and adjacent storage of liquids. Main aisles shall be a minimum of 8 ft (2.4 m) wide. Where ordinary combustible commodities are stored in the same area as liquids in containers, the minimum distance between the two types of storage shall be 8 ft (2.4 m).

In order to reduce the potential of serious fire hazards in warehouses, it is necessary to provide adequate separation of containers by aisles. Generally, minimum aisle widths are obtained by providing sufficient space to operate a fork-lift truck. However, the more important criteria are the fire hazard characteristics of the stored material. Aisle widths should permit adequate access for both detection of leaking containers and the use of portable fire extinguishers and hose streams. Piles must be arranged so that: (a) there are at least 4 ft (1.2 m) aisles between the piles, (b) no container in a pile is more than 12 ft (3.6 m) from an aisle, (c) 4 ft (1.2 m) wide aisles are provided between rack storage and adjacent rows of racks, (d) main aisles are a minimum of 8 ft (2.4 m) wide, and (e) at least 8 ft (2.4 m) aisles are provided between storage piles of flammable and combustible liquids and storage piles of other ordinary combustible commodities. It is also important that aisles be wide enough to safely operate lift trucks without tipping over, puncturing, or otherwise damaging flammable liquid containers.

4-5.6.7 Mixed Storage. Liquids shall not be stored in the same pile or in the same rack sections as ordinary combustible commodities. Where liquids are packaged together with ordinary combustibles, as in kits, the storage shall be considered on the basis of whichever commodity predominates. When two or more classes of liquids are stored in a single pile or single rack section, the maximum quantities permitted in the pile or rack section shall be the smallest of the two or more separate maximum quantities, and the height of storage permitted in that pile or rack section shall be the least of the two or more separate heights. The maximum total quantities permitted shall be limited to the sum of proportional amounts that each class of liquid present bears to the maximum total permitted for its respective class. The sum of proportional amounts shall not exceed 100 percent.

The requirements for storing two or more classes of flammable liquids in a

single pile or rack are the same as the requirements for the mixed storage in cutoff rooms and attached buildings. (See 4-4.2.9 and comments.)

4-5.7 Liquid Warehouses. *(See 1-2 Definitions.)*

4-5.7.1 Liquid warehouses shall be separate, detached buildings or shall be separated from other type occupancies by standard 4 hr fire walls, with communicating openings protected on each side of the wall with automatic-closing listed 3 hr (A) fire doors. Fire doors shall be installed in accordance with NFPA 80, *Standard for Fire Doors and Windows.*

Liquid warehouses are required to meet more restrictive construction standards than those required for a general purpose warehouse. Liquid warehouses are not allowed to be separated from adjoining areas by a 2-hr fire partition. Such separation must be by a 4-hr fire wall. (See comments under 4-5.6.1.) The requirement for two fire doors on the fire wall remains because of the potentially more severe hazard connected with liquid warehouses.

4-5.7.2 If the warehouse building is located more than 10 ft (3.05 m) but less than 50 ft (15.2 m) from an important building or line of adjoining property than can be built upon, the exposing wall shall have a fire resistance rating of at least 2 hr with each opening protected with a listed 1½ hr (D) fire door.

In addition, the liquid warehouse must have those exterior walls located between 10 to 50 ft (3 to 15 m) from an important building or the property line of an adjoining property that can be built upon, and constructed of at least 2-hr fire resistance design. Each opening must be protected by a listed 1½ hr (D) fire door. A (D) fire door is designed for installation in exterior walls which are subject to severe fire exposure from outside the building. (See comment under 4-4.2.2.)

4-5.7.3 If the warehouse is located 10 ft (3.05 m) or less from an important building or line of adjoining property that can be built upon, the exposing wall shall have a fire resistance rating of 4 hr with each opening protected with a listed 3 hr (A) fire door.

The (A) rated door is designed for use in walls separating buildings or walls which divide a building into fire areas. (See comment under 4-5.6.1.) Note that the *Code* is silent on warehouses located beyond 50 ft (15 m) from an important building or the adjoining property that can be built upon. This implies no construction requirements.

4-5.7.4 An attached warehouse, having communicating openings in the required 4 hr fire wall separation from the adjacent building area, shall have these openings protected by:

(a) Normally closed listed 3 hr (A) fire doors on each side of the wall. These doors may be arranged to stay open during material handling operations, only if the doors are designed to close automatically in a fire emergency by provision of listed closure devices.

(b) Noncombustible liquidtight, raised sills or ramps, at least 4 in. (101.6 mm) in height, or other design features to prevent flow of liquids to the adjoining area.

4-5.7.5 Fire doors shall be installed in accordance with NFPA 80, *Standard for Fire Doors and Windows.*

4-5.7.6 The total quantity of liquids within a liquid warehouse shall not be restricted. The maximum pile heights and maximum quantity per pile, arranged as palletized and/or solid pile storage, shall comply with Table 4-4.2.7, if unprotected, or Table 4-6.1(a) if protected in accordance with Section 4-6. The storage heights of containers on protected racks shall comply with Table 4-6.1(b) as applicable.

Exception: An unprotected liquid warehouse located a minimum of 100 ft (30.5 m) from exposed buildings or adjoining property that can be built upon is not required to conform to Table 4-4.2.7, if there is protection for exposures. Where protection for exposures is not provided, a minimum 200 ft (60.9 m) distance is required.

Some confusion develops when Tables 4-4.2.7, 4-6.1(a), and 4-6.1(b) are consulted. Each of these tables has a maximum total quantity column. It must be remembered that liquid warehouses are *not* governed by the maximum quantities indicated in these tables. Liquid warehouses can store unrestricted total amounts. The tables are to be used for regulation of pile heights and maximum quantities per pile.

Storage requirements for liquid warehouses are similar to those for general purpose warehouses. However, the *Code* specifies additional requirements here for empty pallet storage, stability, portable tank design, and storage height relative to roof members.

The Exception to 4-5.7.6 is new to the 1981 edition of the *Code*. The Technical Committee, in considering a public proposal, agreed that the benefits to be derived from the protection of a liquid warehouse may be minimal. For example, bulk supplies of inexpensive, readily replaced flammable liquids may be stored in a well-detached, unprotected liquid warehouse with weekly supplies transferred to a properly protected cutoff liquid storage area. The detached, unprotected facility would pose a minimal exposure to people and important buildings.

4-5.7.7 Class I liquids shall not be permitted in the basement areas of liquid warehouses. Class II and Class IIIA liquids may be stored in basements provided that automatic sprinkler protection and other fire protection facilities are provided in accordance with Section 4-6.

CONTAINER AND PORTABLE TANK STORAGE 127

See comment following 4-4.2.12.

4-5.7.8 Limited amounts of combustible commodities, as defined in the scope of NFPA 231, *Indoor General Storage*, and NFPA 231C, *Rack Storage of Materials*, may be stored in liquid warehouses if protection is provided in accordance with Section 4-6, and the ordinary combustibles, other than those used for packaging the liquids, are separated a minimum of 8 ft (2.4 m) horizontally, by aisles or open racks, from the liquids in storage.

The "limited" amounts bring the judgment of the authority having jurisdiction into the picture.

4-5.7.9 Empty or idle combustible pallet storage shall be limited to a maximum pile size of 2500 sq ft (232 m^2) and to a maximum storage height of 6 ft (1.8 m). Idle pallet storage shall be separated from liquids by at least 8 ft (2.4 m) wide aisleways. However, pallet storage in accordance with NFPA 231, *Standard on Indoor General Storage*, shall be acceptable.

This restriction is intended to place a reasonable upper limit on fire loading.

4-5.7.10 Containers in piles shall be separated by pallets or dunnage to provide stability and to prevent excessive stress on container walls. Portable tanks stored over one tier high shall be designed to nest securely, without dunnage. (*See NFPA 386, Standard on Portable Shipping Tanks, for information on portable tank design*.) Materials handling equipment shall be suitable to handle containers and tanks safely at the upper tier level.

4-5.7.11 No container or portable tank shall be stored closer than 36 in. (914.4 mm) to the nearest beam, chord, girder or other roof member in an unprotected warehouse.

4-5.7.12 Solid pile and palletized storage shall be arranged so that piles are separated from each other by at least 4 ft (1.2 m). Aisles shall be provided so that no container or tank is more than 12 ft (3.6 m) from an aisle. Where storage on racks exists as permitted in this code, a minimum 4-ft (1.2-m) wide aisle shall be provided between adjacent rows of racks and any adjacent storage of liquids. Main aisles shall be a minimum of 8 ft (2.4 m) wide, and access shall be maintained to all doors required for egress.

4-5.7.13 Mixed Storage. When two or more classes of liquids are stored in a single pile, the maximum quantity permitted in that pile shall be the smallest of the two or more separate maximum quantities and the heights of storage permitted in that pile shall be the least of the two or more separate heights as given in Tables 4-4.2.7 or 4-6.1(a) as applicable. When two or more classes of liquids are stored in the same racks as permitted in this code, the maximum height of storage permitted shall be the least of the two or more separate heights given in Table 4-6.1(b).

There is no ceiling on the total amount of liquid that can be stored. Therefore, one need not be concerned with proportional amounts of different classes totalling 100 percent, as was the case with other types of indoor storage.

When it comes to pile storage, however, there is a restriction on mixed storage, both as to total pile quantity and pile height. In essence, the pile takes on the identity of the most hazardous class stored therein, and the total amount in the pile cannot exceed the amount specified for that class. The same principle applies to the pile height.

4-6 Protection Requirements for Protected Storage of Liquids.

4-6.1 Containers and portable tanks storing flammable and combustible liquids may be stored in the quantities and arrangements specified in Tables 4-6.1(a) and 4-6.1(b) provided the storage is protected in accordance with 4-6.2 and 4-6.5, as applicable.

> The referenced Tables are to be used for flammable and combustible liquid storage in *protected* areas. For *unprotected* storage arrangements, see Table 4-4.2.7.

4-6.1.1 Other quantities and arrangements may be used where suitably protected and approved by the authority having jurisdiction.

4-6.2 Where automatic sprinklers are used, they shall be installed in accordance with NFPA 13, *Standard for the Installation or Sprinkler Systems*, and approved by the authority having jurisdiction. (*For additional information, see Appendix C.*)

> Note that Table C-4-6.2(a) in Appendix C gives no advice on protection for 55 gal (208.2 L) containers of Class IB, IC, and II liquids stored more than one high, although such storage is permitted in Table 4-5.1(a).

4-6.2.1 Other systems such as automatic foam-water systems, automatic water-spray systems, or other combinations of systems may be considered acceptable if approved by the authority having jurisdiction. (*For additional information, see Appendix C.*)

4-6.3 Racks storing Class I or Class II liquids shall be either single-row or double-row as described in NFPA 231C, *Rack Storage of Materials*.

4-6.4 Ordinary combustibles other than those used for packaging the liquids shall not be stored in the same rack section as liquids, and shall be separated a minimum of 8 ft (2.4 m) horizontally, by aisles or open racks, from liquids stored in racks.

> When applying the provisions of Tables 4-6.1(a) and 4-6.1(b), it is important

Table 4-6.1(a) Storage Arrangements for Protected Palletized or Solid Pile Storage of Liquids in Containers and Portable Tanks

Class	Storage Level	Max. Stge. Height (ft.) Containers	Max. Stge. Height (ft.) Port. Tanks	Max. Quantity per Pile (gal.) Containers	Max. Quantity per Pile (gal.) Port. Tanks	Max. Quantity (gal.) Containers	Max. Quantity (gal.) Port. Tanks
IA	Ground Floor	5	—	3,000	—	12,000	—
	Upper Floors	5	—	2,000	—	8,000	—
	Basements	—Not Permitted—		—	—	—	—
IB	Ground Floor	6½	7	5,000	20,000	15,000	40,000
	Upper Floors	6½	7	3,000	10,000	12,000	20,000
	Basements	—Not Permitted—		—	—	—	—
IC	Ground Floor	*6½	7	5,000	20,000	15,000	40,000
	Upper Floors	*6½	7	3,000	10,000	12,000	20,000
	Basements	—Not Permitted—		—	—	—	—
II	Ground Floor	10	14	10,000	40,000	25,000	80,000
	Upper Floors	10	14	10,000	40,000	25,000	80,000
	Basements	5	7	7,500	20,000	7,500	20,000
III	Ground Floor	20	14	15,000	60,000	50,000	100,000
	Upper Floors	20	14	15,000	60,000	50,000	100,000
	Basements	10	7	10,000	20,000	25,000	40,000

SI Units: 1 ft = 0.3048 m; 1 gal = 3.785 L.

*These height limitations may be increased to 10 ft for containers of 5 gal or less in capacity.

Note: See Section 4-6 for protection requirements as applicable to this type of storage.

to consider each restricting limitation in succession so that none will be overlooked. It is also necessary to make sure that the differing limits for container or portable tanks are observed.

Table 4-6.1(b) **Storage Arrangements for Protected Rack Storage of Liquids in Containers**

Class	Type Rack	Storage Level	Max. Stge. Height (ft.) Containers	Max. Quantity (gal.) Containers
IA	Double Row or Single Row	Ground Floor Upper Floor Basements	25' 15' Not Permitted	7,500 4,500 —
IB	Double Row or Single Row	Ground Floor Upper Floor Basements	25' 15' Not Permitted	15,000 9,000 —
IC				
II	Double Row or Single Row	Ground Floor Upper Floor Basements	25' 25' 15'	24,000 24,000 9,000
III	Multi-Row Double Row or Single Row	Ground Floor Upper Floor Basements	40' 20' 20'	48,000 48,000 24,000

SI Units: 1 ft = 0.3048 m; 1 gal = 3.785 L.

Note: See Section 4-6 for protection requirements as applicable to this type of storage.

Note that Tables 4-6.1(a) and 4-6.1(b) are the storage arrangements for *protected* storage of flammable and combustible liquids in containers and portable tanks. For *unprotected* storage arrangements, see Table 4-4.2.7.

4-6.5 In-rack sprinklers shall be installed in accordance with the provisions of NFPA 231C, *Rack Storage of Materials*, except as modified by 4-6.2. Alternate lines of in-rack sprinklers shall be staggered. Multiple levels of in-rack sprinkler heads shall be provided with water shields unless otherwise separated by horizontal barriers, or unless the sprinkler heads are listed for such installations.

4-7 Fire Control.

4-7.1 Suitable fire extinguishers or preconnected hose lines, either 1½ in. (38.1 mm) lined or 1 in. (25.4 mm) hard rubber, shall be provided where liquids are stored. Where 1½ in. (38.1 mm) fire hose is used, it shall be installed in accordance with NFPA 14, *Standard for the Installation of Standpipe and Hose Systems*.

Portable fire extinguishers and preconnected hose lines are provided for first aid fire fighting and fire control. Combination (adjustable) nozzles capable of both spray (fog) and straight stream operation are preferred for this type of service. Although not covered in NFPA 14,[8] 1-in. hard rubber hose line installations should follow the principles in that standard.

4-7.1.1 At least one portable fire extinguisher having a rating of not less than 20-B shall be located outside of, but not more than 10 ft (3 m) from, the door opening into any separate inside storage area.

The 1981 edition of the *Code* changed the extinguisher requirement from 10-B to 20-B, based on a public proposal. NFPA 10[9] requires 20-B and 40-B rated extinguishers for extra hazardous locations depending on travel distance.

4-7.1.2 At least one portable fire extinguisher having a rating of not less than 20-B shall be located not less than 10 ft (3 m), nor more than 50 ft (15.2 m), from any Class I or Class II liquid storage area located outside of a separate inside storage area.

The reason for requiring that portable fire extinguishers be located a distance away from the storage room is that fires in Class I and Class II flammable liquids are likely to escalate rapidly. If the fire extinguisher is too close to the storage area, it will be impossible to get to the extinguisher once the fire has started.

4-7.1.3 In protected general purpose and liquid warehouses, hand hose lines shall be provided in sufficient number to reach all liquid storage areas.

It is the intent of this paragraph to require that the hose stream (not the hose

itself) is able to be applied to all areas and sides of the flammable liquid storage area. The hose itself need not reach into every area if the water discharge is not obstructed.

4-7.1.4 The water supply shall be sufficient to meet the fixed fire protection demand, plus a total of at least 500 gal (1892.5 L) per minute for inside and outside hose lines. (*See C-4-6.2.*)

> This water supply requirement is intended to be a minimum *total* water supply requirement rather than a water supply requirement for each hose line.

4-7.2 Control of Ignition Sources. Precautions shall be taken to prevent the ignition of flammable vapors. Sources of ignition include but are not limited to: open flames; lightning; smoking; cutting and welding; hot surfaces; frictional heat; static, electrical and mechanical sparks; spontaneous ignition, including heat-producing chemical reactions; and radiant heat.

> The listed sources of ignition are not intended to be all inclusive, but are included as examples of what may be considered sources of ignition for purposes of enforcing the provisions of this *Code*.

4-7.3 Dispensing of Class I and Class II liquids in general purpose or liquid warehouses shall not be permitted unless the dispensing area is suitably cut off from other ordinary combustible or liquid storage areas, as specified in Section 4-4, and otherwise conforms with the applicable provisions of Section 4-4.

> Because of the greater likelihood of fire occurrence when dispensing or transferring flammable liquids, the *Code* prohibits such activity from being conducted in the same areas where there are large quantities of flammable liquid storage.

4-7.4 Materials with a water reactivity degree of 2 or higher as outlined in NFPA 704, *Standard System for the Identification of the Fire Hazards of Materials*, shall not be stored in the same area with other liquids.

> Many flammable and combustible liquid storage areas are protected by automatic sprinkler and water spray systems and hose lines. Consequently, any storage of water reactive material in the flammable liquid storage area creates an unreasonable hazard and risk. The NFPA 704 identification system[10] is of particular value to responding fire service personnel in that it provides them with immediate information about the hazards involved.

4-8 Outdoor Storage.

> Requirements for outdoor storage of flammable and combustible liquids are often very similar to the requirements for inside storage. The major difference

CONTAINER AND PORTABLE TANK STORAGE   133

is that while outdoor storage provides natural ventilation, the fire exposure problem is more critical from and to buildings, flammable liquid tanks, or ordinary combustible materials. The *Code* recognizes that there is a significantly reduced fire hazard potential with smaller quantities of flammable and combustible liquid than there would be for larger quantities.

Outdoor storage presents security problems. On the other hand, storing flammable and combustible liquids outdoors reduces the structural fire hazard. Generally, a fire in an outdoor storage facility is more readily controlled than one within a structure.

4-8.1   Outdoor storage of liquids in containers and portable tanks shall be in accordance with Table 4-8, as qualified by 4-8.1.1 through 4-8.1.4 and 4-8.2, 4-8.3, and 4-8.4.

4-8.1.1   When two or more classes of materials are stored in a single pile, the maximum gallonage in that pile shall be the smallest of the two or more separate gallonages.

This means that the pile takes on the identity of the more hazardous class, and the total amount in the pile cannot exceed that allowed for that class.

4-8.1.2   No container or portable tank in a pile shall be more than 200 ft (60.9 m) from a 12 ft (3.65 m) wide access way to permit approach of fire control apparatus under all weather conditions.

The phrase "under all weather conditions" was added to the 1981 edition of the *Code* as a result of a public proposal. The proponent contended that dirt access roads might be impassable in wet weather, and the Committee agreed. This addition addresses the problem.

4-8.1.3   The distances listed in Table 4-8 apply to properties that have protection for exposures as defined. If there are exposures, and such protection for exposures does not exist, the distances in column 4 shall be doubled.

4-8.1.4   When total quantity stored does not exceed 50 percent of maximum per pile, the distances in columns 4 and 5 may be reduced 50 percent, but to not less than 3 ft (0.91 m).

The following three problems illustrate how Table 4-8 is applied:

1. There are 1,700 gal (6434.5 L) of diesel fuel (Class II) in drums stored outside a warehouse. How many gal of gasoline (Class IB) can the plant store in the same pile?

Solution: The Class IB storage is the more hazardous, so provisions gover-

Table 4-8 Outdoor Liquid Storage in Containers and Portable Tanks

	1		2		3	4	5
Class	Container Storage-Max. per Pile Gallons (1) (4)	Height (ft)	Portable Tank Storage Max. per Pile Gallons (1) (4)	Height (ft)	Distance Between Piles or Racks (ft)	Distance to Property Line That Can Be Built Upon (ft)(2)(3)	Distance to Street, Alley, or a Public Way (ft) (3)
IA	1,100	10	2,200	7	5	50	10
IB	2,200	12	4,400	14	5	50	10
IC	4,400	12	8,800	14	5	50	10
II	8,800	12	17,600	14	5	25	5
III	22,000	18	44,000	14	5	10	5

SI Units: 1 ft = 0.3048 m; 1 gal = 3.785 L.

Notes: (1) See 4-8.1.1 regarding mixed class storage.
(2) See 4-8.1.3 regarding protection for exposures.
(3) See 4-8.1.4 for smaller pile sizes.
(4) For storage in racks, the quantity limits per pile do not apply, but the rack arrangement shall be limited to a maximum of 50 feet in length and two rows or 9 feet in depth.

ning IB pile storage are applicable. The maximum allowed per pile is 2,200 gal (8327 L). Since the plant already has 1,700 gal (6434.5 L) of Class II, only 500 gal (1892.5 L) of gasoline can be stored in the same pile.

2. A manufacturing plant has no protection for exposures. Several piles of drums containing gasoline (Class IB) are located 125 ft (38.1 m) from the nearest property line that can be built upon. An industrial plant on the adjacent property is in the process of acquiring its own fire brigade. What will be the new distance requirement for the manufacturing plant's gasoline containers?

Solution: When there is no protection for exposures, the distance specified in Column 4 must be doubled. Thus, 100 ft (30.48 m) is needed. The plant in question has its storage 125 ft (38.1 m) away, so there is no *Code* violation. When the protection for exposures is in effect, the plant can move its storage to a 50 ft (15.2 m) distance from its neighbor.

3. How close to adjacent property which can be built upon, and how close to a nearby street, is the manufacturing plant allowed to place 2,300 gal (8705.5 L) of kerosene (Class II) in drums? Assume there is no protection for exposures.

Solution: With no protection for exposures, the distance given in Column 4 must be doubled to 50 ft (15.2 m). The plant is allowed to store 8,800 gal (33 308 L), but has only stored, 2,200 gal (8327 L). This is substantially less than 50 percent of the allowable amount, so the plant can reduce the distance specified in Column 4 by 50 percent. Therefore, storage can be placed 25 ft (7.6 m) from the property line that can be built upon.

Because there is less than 50 percent of the allowable amount, the plant may reduce the distance in Column 5 to 50 percent, but to no less than 3 ft (0.9144 m). Therefore, the plant's storage must be at least 3 ft (0.9144 m) from the nearest street.

4-8.2 A maximum of 1,100 gal (4163.5 L) of liquids in closed containers and portable tanks may be stored adjacent to a building located on the same premises and under the same management provided that:

(a) The building is limited to a one-story building of fire-resistive or noncombustible construction and is devoted principally to the storage and handling of liquids, or

(b) The building has an exterior wall with a fire resistance rating of not less than 2 hrs and having no opening to above grade areas within 10 ft (3.05 m) horizontally of such storage and no openings to below grade areas within 50 ft (15.24 m) horizontally of such storage.

4-8.2.1 The quantity of liquids stored adjacent to a building protected in accordance with 4-8.2(b) may exceed that permitted in 4-8.2, provided the maximum quantity per pile does not exceed 1,100 gal (4163.5 L) and each pile is separated by a 10 ft (3.05 m) minimum clear space along the common wall.

4-8.2.2 Where the quantity stored exceeds the 1,100 gal (4163.5 L) permitted adjacent to the building given in 4-8.2(a), or the provisions of 4-8.2(b) cannot be met, a minimum distance in accordance with column 4 of Table 4-8 shall be maintained between buildings and nearest container or portable tank.

4-8.3 The storage area shall be graded in a manner to divert possible spills away from buildings or other exposures or shall be surrounded by a curb at least 6 in. (152.4 mm) high. When curbs are used, provisions shall be made for draining of accumulations of ground or rain water or spills of liquids. Drains shall terminate at a safe location and shall be accessible to operation under fire conditions.

4-8.4 Storage area shall be protected against tampering or trespassers where necessary and shall be kept free of weeds, debris and other combustible materials not necessary to the storage.

In addition to the requirements of this *Code* relative to outdoor storage, one must be aware of environmental requirements, particularly relating to the pollution of waterways. Additional dikes, curbing, or drainage systems may have to be provided in order to control any spill that may occur. Drainage or collection systems must be assessed from a fire protection standpoint to eliminate a fire hazard in areas where the drained flammable liquids are collected, no matter how remote from the storage.

REFERENCES CITED BY *CODE*

(These publications comprise a part of the requirements to the extent called for by the **Code.***)*

Department of Transportation, *Code of Federal Regulations,* Chapter 1, Title 49.

NFPA 13, *Standard for the Installation of Sprinkler Systems,* NFPA, Boston, 1980.

NFPA 45, *Standard on Fire Protection for Laboratories Using Chemicals,* NFPA, Boston, 1975.

NFPA 386, *Standard for Portable Shipping Tanks,* NFPA, Boston, 1979.

NFPA 70, *National Electrical Code,* NFPA, Boston, 1981.

NFPA 80, *Standard for Fire Doors and Windows,* NFPA, Boston, 1979.

NFPA 91, *Standard for the Installation of Blower and Exhaust Systems for Dust, Stock and Vapor Removal or Converting,* NFPA, Boston, 1973.

NFPA 231, *Standard for Indoor General Storage,* NFPA, Boston, 1979.

NFPA 231C, *Rack Storage of Materials,* NFPA, Boston, 1980.

NFPA 251, *Standard Methods of Fire Tests of Building Construction and Materials,* NFPA, Boston, 1979.

NFPA 704, *Identification of the Fire Hazards of Materials,* NFPA, Boston, 1980.

NFPA 68, *Guide for Explosion Venting,* NFPA, Boston, 1978.

ASTM D92-78, *Test for Flash and Fire Points by Cleveland Open Cup,* American Society for Testing and Materials, Philadelphia, 1978.

ASTM E152-78, *Fire Tests of Door Assemblies,* American Society for Testing and Materials, Philadelphia, 1978.

REFERENCES CITED IN COMMENTARY

[1] ANSI/ASTM D3435-78, *Specifications for Plastic Containers for Petroleum Products,* American Society for Testing and Materials, Philadelphia, 1978.

[2] ASTM E152-78, *Fire Tests of Door Assemblies,* American Society for Testing and Materials, Philadelphia, 1978.

[3] NFPA 252, *Standard Methods of Fire Tests of Door Assemblies,* NFPA, Boston, 1979.

[4] NFPA 251, *Standard Methods of Fire Tests of Building Construction and Materials,* NFPA, Boston, 1979.

[5] ASTM E119-79, *Fire Tests of Building Construction and Materials,* American Society for Testing and Materials, Philadelphia, 1979.

[6] NFPA 325M, *Fire Hazard Properties of Flammable Liquids, Gases, and Volatile Solids,* NFPA, Boston, 1977.

[7] NFPA 13, *Installation of Sprinkler Systems,* NFPA, Boston, 1980.

[8] NFPA 14, *Standpipe and Hose Systems,* NFPA, Boston, 1978.

[9] NFPA 10, *Portable Fire Extinguishers,* NFPA, Boston, 1978.

[10] NFPA 704, *Identification of the Fire Hazards of Materials,* NFPA, Boston, 1980.

5

Industrial Plants

5-1 Scope.

5-1.1 This chapter shall apply to those industrial plants where (1) the use of liquids is incidental to the principal business (*see Section 5-2*), or (2) where liquids are handled or used only in unit physical operations such as mixing, drying, evaporating, filtering, distillation, and similar operations which do not involve chemical reaction (*see Section 5-3*). This chapter shall not apply to chemical plants, refineries or distilleries, as defined, which are covered in Chapter 9, Refineries, Chemical Plants and Distilleries.

The reason for three different chapters — 5, 8, and 9 — is that the Technical Committee feels there are three types of hazards which, despite some overlapping, require different philosophical approaches. In this chapter, the hazards involve the likelihood of vapor release, since large quantities of liquids may be exposed to the air, as in mixing, drying, or filtering. There is a consequent potential for high rates of evolution of hazardous material, in the event that heated liquids or their vapors are accidentally released.

In Chapter 8, Processing Plants, the same potentials may exist as in this chapter. However, there is the added risk that vapor or liquid release may be triggered by the heat of uncontrolled chemical reactions. This potential may or may not involve the use of reactive liquids.

In Chapter 9, Refineries, Chemical Plants, and Distilleries, the latter will often have only the hazards associated with Chapter 5, except perhaps for the association of low hazard fermentation. Refineries (oil) and chemical plants will have the hazards associated with both Chapters 5 and 8. The reasons for putting these in a separate chapter are obvious.

Distilleries, primarily for ethyl alcohol, are closely regulated by governmental taxing authorities. This usually results in large storages indoors. However, precautions against theft and contamination are such that the hazards of accidental spill or ignition are greatly reduced. Alcohol is water soluble; thus, sprinkler systems and conventional fire fighting with water have added effectiveness because cooling is accompanied by dilution. One hundred-proof alcohol is 50 percent water, with a flash point of 75°F (23.9°C). If 4 gal (15.14 L) of water are added to 1 gal (3.785 L) of the 100-proof material, the flash point goes up to 120°F (48.9°C). Another feature of distilleries is that whiskey being aged is stored in wooden barrels. In case of fire, the metal hoops expand and the barrels leak, eliminating internal buildup of pressure and preventing explosions.

Oil refineries present special problems. Except for hydrogenations, most chemical processes in refineries involve the heating of material to crack or refine it. Both are heat absorbing (endothermic) reactions. Because high temperatures are involved, fired equipment is common and loss prevention becomes a matter of proper operation, good maintenance, and the design, metallurgy, and location of equipment and piping. Since major operations are in equipment located outside buildings, the detailed requirements in Chapters 5 and 8 are mostly inapplicable.

Chemical plants differ from processing plants in that they are large and integrated. As with refineries, most large-scale use of liquids takes place outdoors. However, work inside buildings may be necessary where toxicity or product purity (as with pharmaceuticals) is involved. Most "large integrated" plants have a safety organization familiar with hazard protection.

5-1.2 Where portions of such plants involve chemical reactions such as oxidation, reduction, halogenation, hydrogenation, alkylation, polymerization, and other chemical processes, those portions of the plant shall be in accordance with Chapter 8, Processing Plants.

An example would be the reaction with ethylene oxide of a small amount of the product of a plant whose main operation was the solvent extraction of vegetable oil from an oil seed.

5-2 Incidental Storage or Use of Liquids.

5-2.1 Section 5-2 shall be applicable to those portions of an industrial plant where the use and handling of liquids is only incidental to the principal business, such as automobile assembly, construction of electronic equipment, furniture manufacturing or other similar activities.

This is intended to cover liquids in the form of paint thinners, cleaning solvents, janitorial aids, etc.

5-2.2 Liquids shall be stored in tanks or closed containers.

5-2.2.1 Except as provided in 5-2.2.2 and 5-2.2.3, all storage shall comply with Chapter 4, Container Storage.

This removes some of the limitations in Chapter 4.

5-2.2.2 The quantity of liquid that may be located outside of an inside storage room or storage cabinet or in any one fire area of a building shall not exceed the greater of that given in (a) or (b), (c) and (d) below:

(a) A supply for one day, or

(b) 25 gal (94.6 L) of Class IA liquids in containers, and

(c) 120 gal (454.2 L) of Class IB, IC, II or III liquids in containers, and

(d) One portable tank not exceeding 660 gal (2498.1 L) of Class IB, IC, Class II or Class III liquids.

In connection with (a) and (c), steel drums could be used. The violent rupture of such sealed drums exposed to fire is a well-known phenomenon. Plastic drums are self-venting. Portable tanks meeting DOT requirements have such vents.

5-2.2.3 Where large quantities of liquids are necessary, storage may be in tanks, which shall comply with the applicable requirements of Chapter 2, Tank Storage, and Sections 5-3, 5-4, 5-5, 5-6, 5-7 and 5-8.

Use of a tank, pump, and piping eliminates the hazard of frequent transfer accompanying small containers.

5-2.3 Areas in which liquids are transferred from one tank or container to another container shall be separated from other operations in the building by adequate distance or by construction having adequate fire resistance. Drainage or other means shall be provided to control spills. Adequate natural or mechanical ventilation shall be provided. NFPA 91, *Standard for the Installation of Blower and Exhaust Systems for Dust, Stock and Vapor Removal or Conveying*, provides information on the design and installation of mechanical ventilation.

Because this paragraph relies on the word "adequate," the requirements will all be the result of judgment. For example, where distance is concerned, a leak of 1 gal per min (0.063L/s) of gasoline will feed a fire having a diameter of about 8 ft (2.4 m). While no one could stand closer than 20 to 25 ft (6.1 m to 7.62 m) from such a fire, a simple metal partition would stop the radiant heat. Drainage would not let the pool spread to its 8 ft (2.4 m) diameter, but a curb would. If the spill did not ignite, ventilation, in accordance with 5-3.5, would solve the problem. However, vapors not rich enough to be flammable, but well above the limits permitted on the basis of toxicity, could spread farther. The best solution would be to have a storage area (as in Chapter 4) in which to make the transfer into safety cans.

5-2.4 Handling Liquids at Point of Final Use.

5-2.4.1 Class I and Class II liquids shall be kept in covered containers when not actually in use.

Note that the word is "covered," and not "closed" or "sealed." This permits the use of small containers into which, for example, metal parts could be dipped for rinsing off oil.

5-2.4.2 Where liquids are used or handled, except in closed containers, means shall be provided to dispose promptly and safely of leakage or spills.

Here the word is "closed." The requirement may be met by providing approved waste cans.

5-2.4.3 Class I liquids may be used only where there are no open flames or other sources of ignition within the possible path of vapor travel.

Paragraph 5-3.5.1 could be consulted here. However, if health standards are met, the fire hazard is small.

5-2.4.4 Class I and Class II liquids shall be drawn from or transferred into vessels, containers, or portable tanks within a building only from (1) original shipping containers with a capacity of 5 gal (18.92 L) or less, or (2) from safety cans, or (3) through a closed piping system, or (4) from a portable tank or container by means of a device drawing through an opening in the top of the tank or container, or (5) by gravity through a listed self-closing valve or self-closing faucet.

5-2.4.5 Transferring liquids by means of pressurizing the container with air is prohibited. Transferring liquids by pressure of inert gas is permitted only if controls, including pressure relief devices, are provided to limit the pressure so it cannot exceed the design pressure of the vessel, tank or container.

Since the drums ordinarily used for liquids are not qualified as pressure containers and no containers made to UL specifications are so qualified, this paragraph essentially limits the use of gas pressure to containers built to the ASME *Pressure Vessel Code*[1] or to ones qualified as pressure containers under DOT regulations. Note that by definition, containers have a capacity of 60 gal (227.1 L) or less. The reason for banning the use of air, even though the vapor space above liquid in the container may already contain air, is that the compressed air will raise the UFL (upper flammable limit) of a mixture of vapor with air. Also, if we assume that a combustion explosion raises the pressure about 8 times and that normal atmospheric pressure is 15 psia, then an explosion in an unpressurized drum would raise the internal pressure to 120 psia or 120 − 15 = 105 psig. Five lbs air pressure would give (15 + 5) × 8 or 160 − 15, that is, 145 lbs pressure and a correspondingly more violent explosion. The highest test pressure required by DOT is 125 psig for a specification 5-F container. Tests required on others may be 15, 20, 30, 40, 60, 80, or 100 psig. Obviously, if inert gas pressure is to be used, the specification for the specific type of container must be consulted to ensure compliance with the last sentence.

INDUSTRIAL PLANTS 143

5-3 Unit Physical Operations.

5-3.1 Section 5-3 shall be applicable in those portions of industrial plants where liquids are handled or used in unit physical operations such as mixing, drying, evaporating, filtering, distillation, and similar operations which do not involve chemical change. Examples are plants compounding cosmetics, pharmaceuticals, solvents, cleaning fluids, insecticides and similar types of activities.

See the discussion following 5-1.1.

5-3.2 Industrial plants shall be located so that each building or unit of equipment is accessible from at least one side for fire fighting and fire control purposes. Buildings shall be located with respect to lines of adjoining property which may be built upon as set forth in 8-2.1 and 8-2.1.1, except that the blank wall referred to in 8-2.1.1 shall have a fire resistance rating of at least 2 hr.

The concept here is that equipment containing a liquid may be either in the open or housed, and that the distance to a property line should be measured from the container. If housed, the building wall and the required drainage serve the function of the dike. If not housed, although not specifically required, there should be dikes or drainage in accordance with Chapter 2.

5-3.3 Areas where unstable liquids are handled or small scale unit chemical processes are carried on shall be separated from the remainder of the plant by a fire wall having a fire resistance rating of not less than 2 hr.

See Figure 5-1.

Figure 5-1. The aftermath of a fatal explosion in a processing vessel within an industrial plant.

This permits *small scale* chemical operations to be carried on without complying with the structural requirements of Chapter 8. For example, the candle factory process of bleaching beeswax by adding small amounts of potassium permanganate to the melted wax.

5-3.4 Drainage.

5-3.4.1 Emergency drainage systems shall be provided to direct flammable or combustible liquid leakage and fire protection water to a safe location. This may require curbs, scuppers, or special drainage systems to control the spread of fire (*see 2-2.3.1*). Appendix A of NFPA 15, *Standard for Water Spray Fixed Systems for Fire Protection*, provides information on such protection.

If scuppers are used, they must not, of course, be in a wall that acts as a dike. The reference to Appendix A of NFPA 15[2] is to a vented ditch which is designed so that it will not be damaged by ignition of vapors, but which is also designed so that any fire in it will be "lazy" because it is starved for air.

5-3.4.2 Emergency drainage systems, if connected to public sewers or discharged into public waterways, shall be equipped with traps or separators.

The use of traps is to ensure that there will be a liquid seal through which fire will not pass. Obviously, the trapped drain must not discharge to a point or system unacceptable to the local or federal authorities. (See 5-3.4.3.)

Separators are useful when a two-phase system — oil and water, for example — is discharged and only one phase needs control. (See Figure 5-2 for drainage system details for areas containing flammable liquids.)

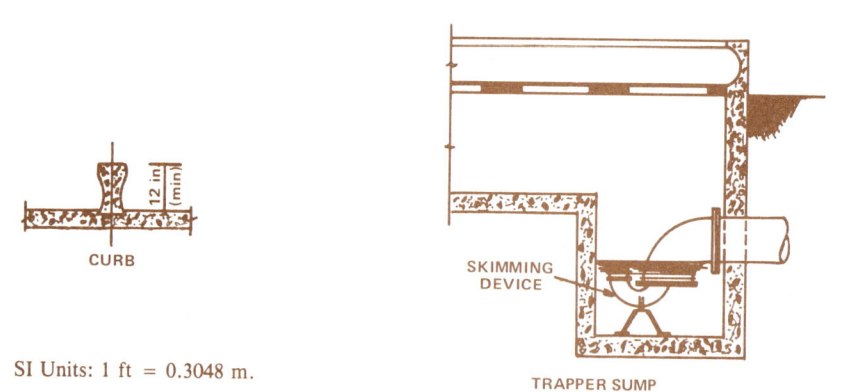

SI Units: 1 ft = 0.3048 m.

Figure 5-2. Drainage system details for areas containing flammable liquids.

5-3.4.3 The industrial plant shall be designed and operated to prevent the normal discharge of flammable or combustible liquids into public waterways, public sewers, or adjoining property.

INDUSTRIAL PLANTS 145

It is occasionally possible to get permission to discharge combustible liquids into public sewers where there is a sewage treatment system capable of handling the discharge.

5-3.5 Ventilation.

5-3.5.1 Areas as defined in 5-3.1 using Class I liquids shall be ventilated at a rate of not less than 1 cu ft (0.028 m³) per min per sq ft (0.0929 m²) of solid floor area. This shall be accomplished by natural or mechanical ventilation with discharge or exhaust to a safe location outside of the building. Provision shall be made for introduction of make-up air in such a manner as not to short circuit the ventilation. Ventilation shall be arranged to include all floor areas or pits where flammable vapors can collect. Local or spot general ventilation may be needed for the control of special fire or health hazards. Such ventilation, if provided, may be utilized for up to 75 percent of the required ventilation. NFPA 91, *Standard for the Installation of Blower and Exhaust Systems for Dust, Stock and Vapor Removal or Conveying*, and NFPA 90A, *Standard for the Installation of Air Conditioning and Ventilating Systems*, of other than residence type, provide information on this subject.

See Figure 5-3.

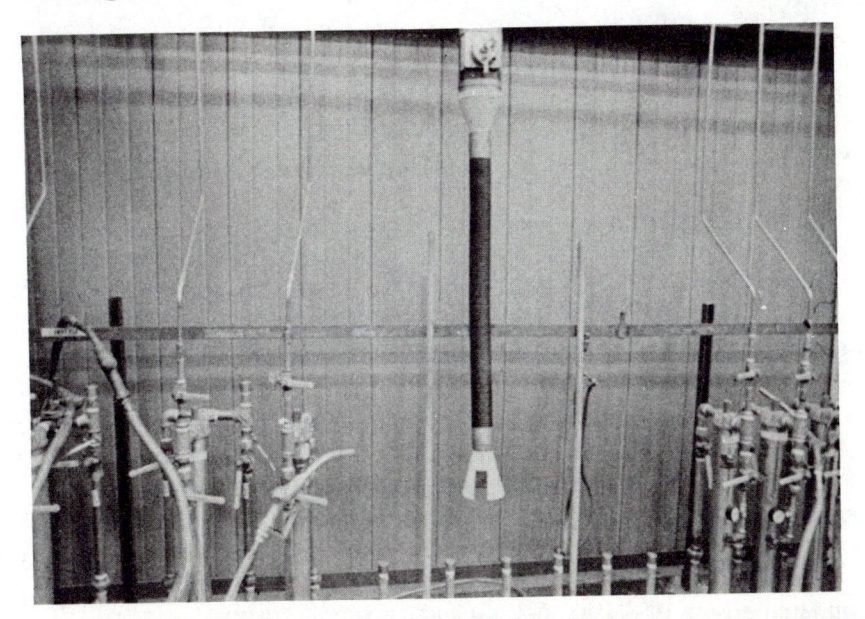

Figure 5-3. Some operations may require local or spot ventilation for control of special fire or health hazards. If provided, local or spot ventilation may be used for up to 75 percent of the required ventilation.

The minimum ventilation rate of 1 cu ft per min per sq ft of solid floor area was developed from the old standard recommendation that areas where low

146   FLAMMABLE AND COMBUSTIBLE LIQUIDS CODE HANDBOOK

flash point flammable liquids were used should be ventilated at the rate of six air changes per hr; that is, one air change every 10 min. However, if a building using liquids has a high ceiling — for example, a foundry with a roof 30 ft (9 m) high — the 6 air changes would involve moving 3 times as much air, with a corresponding loss of heat in the winter, as compared to a room having a 10 ft (3 m) ceiling level. At the boiling point of the liquid the vapors are heavier than air at normal temperature. However, at normal temperatures the vapor is diluted with air and hence not much heavier. Since an imaginary 10 ft (3 m) ceiling is assumed, the requirement, as now written, stipulates the same 6 air changes per hr in the 10 ft (3 m) high space. It is required that special attention be given to low spots.

5-3.5.2 Equipment used in a building and the ventilation of the building shall be designed so as to limit flammable vapor-air mixtures under normal operating conditions to the interior of equipment, and to not more than 5 ft (1.5 m) from equipment which exposes Class I liquids to the air. Examples of such equipment are dispensing stations, open centrifuges, plate and frame filters, open vacuum filters, and surfaces of open equipment.

Since vapors from almost all liquids cause environmentally unacceptable working conditions at concentrations well below the lower flammable limit concentration, this requirement is usually superfluous. Here, as in other instances, 1-1.3 is important.

5-3.6 The storage, transfer and handling of liquids shall comply with Section 8-4 of Chapter 8, Processing Plants.

These items are discussed under Section 8-4.

5-4 Tank Vehicle and Tank Car Loading and Unloading.

5-4.1 Tank vehicle and tank car loading or unloading facilities shall be separated from aboveground tanks, warehouses, other plant buildings or nearest line of adjoining property which can be built upon by a distance of 25 ft (7.62 m) for Class I liquids and 15 ft (4.57 m) for Class II and Class III liquids, measured from the nearest position of any fill stem. Buildings for pumps or shelters for personnel can be a part of the facility. Operations of the facility shall comply with the appropriate portions of Section 6-3 of Chapter 6, Bulk Plants.

The 25 ft (7.62 m) and 15 ft (4.57 m) distances specified here take precedence over those specified in Table 2-6, which relate to separation between a tank and the nearest important building. These items are discussed further under Section 6-3.

5-5 Fire Control.

5-5.1 Portable fire extinguishment and control equipment shall be provided in such quantities and types as are needed for the special hazards of operation and storage. NFPA 10, *Standard for Portable Fire Extinguishers*, provides information as to the suitability of various types of extinguishers and their number and location.

5-5.2 Water shall be available in volume and at adequate pressure to supply water hose streams, foam-producing equipment, automatic sprinklers or water spray systems as the need is indicated by the special hazards of operation, dispensing and storage.

5-5.3 Special extinguishing equipment such as that utilizing foam, inert gas, or dry chemical shall be provided as the need is indicated by the special hazards of operation, dispensing and storage.

5-5.4 Where the need is indicated by special hazards of operation, liquid processing equipment, major piping, and supporting steel shall be protected by approved water spray systems, deluge systems, approved fire resistant coatings, insulation, or any combination of these. NFPA 13, *Standard for the Installation of Sprinkler Systems*, and NFPA 15, *Standard for Water Spray Fixed Systems for Fire Protection*, provide information on this subject.

> **These all refer to the provision of portable and fixed fire protection devices and systems "as the need is indicated by the special hazards of operation," etc. Determining the need should be a joint effort, involving the owner, the local fire authorities, and the insurance carrier, if any. One approach to estimating the size and frequency of potential fires is given in the NFPA publication, *The Systems Approach to Fire Protection*.[3]**

5-5.5 An approved fire alarm system is recommended for prompt notification of fire. Where service is available, it is recommended that a public fire alarm box be located nearby. It may be advisable to connect the plant system with the public system. NFPA 72D, *Standard for the Installation, Maintenance and Use of Proprietary Protective Signaling Systems for Watchman, Fire Alarm and Supervisory Sevice*, provides information on this subject.

5-5.6 All plant fire protection facilities shall be adequately maintained and periodically inspected and tested to make sure they are always in satisfactory operating condition, and they will serve their purpose in time of emergency.

> **The manufacturers' instructions should be followed with respect to maintenance.**

5-6 Sources of Ignition.

5-6.1 Precautions shall be taken to prevent the ignition of flammable vapors. Sources of ignition include but are not limited to: open flames; lightning; smoking; cutting and welding; hot surfaces; frictional heat; static, electrical and mechanical sparks; spontaneous ignition, including heat-producing chemical reactions; and radiant heat.

> Because of the multiplicity of sources of ignition, the best precautions consist of limiting the evolution of flammable vapors and their dispersion into the atmosphere. Naturally, smoking, open flames, cutting and welding, and hot work in general should be controlled whether or not there are flammable vapors present. Things other than flammable vapors can be ignited by accident.

5-6.2 Class I liquids or Class II or Class III liquids at a temperature above their flash points (*see 1-1.3*) shall not be dispensed into metal containers unless the nozzle or fill pipe is in electrical contact with the container. This can be accomplished by maintaining metallic contact during filling, by a bond wire between them, or by other conductive path having an electrical resistance not greater than 10^6 ohms. Bonding is not required where a container is filled through a closed system, or the container is made of glass or other nonconducting material. NFPA 77, *Recommended Practice on Static Electricity*, provides information on static protection; NFPA 78, *Lightning Protection Code*, provides information on lightning protection.

> The subject of ignition by static electricity was discussed under 2-2.7.4. This paragraph makes the point that if two conductors are brought to the same electrical potential by bonding, a spark cannot jump between them. Connecting the two containers individually to ground gives the same result as bonding, but bonding is easier to check. A combination of bonding plus grounding is not needed.

5-7 Electrical Equipment.

5-7.1 This Section, 5-7, shall apply to areas where Class I liquids are stored or handled or where Class II or Class III liquids are stored or handled at a temperature above their flash points (*see 1-1.3.*) For areas where Class II or Class III liquids only are stored or handled at a temperature below their flash points, the electrical equipment may be installed in accordance with provisions of NFPA 70, *National Electrical Code*, for ordinary locations; however, care shall be used in locating electrical apparatus to prevent hot metal from falling into open equipment.

> Incendive hot metal particles from electrical failure would be expected to come from switch gear faults or from failed ballasts on fluorescent lights. There might be sufficient energy in hot filaments from conventional light bulbs to ignite flammable vapors.

5-7.2 All electrical equipment and wiring shall be of a type specified by and shall be installed in accordance with NFPA 70, *National Electrical Code.*

NFPA 70, *National Electrical Code,*® specifies types of wiring and equipment suitable for various locations.[4] The interest of this *Code*, NFPA 30, lies in locations that may be hazardous because of vapors given off by flammable or combustible liquids.

5-7.3 So far as it applies, Table 5-7.3 shall be used to delineate and classify areas for the purpose of installation of electrical equipment under normal circumstances. In the application of classified areas, a classified area shall not extend beyond an unpierced floor, wall, roof or other solid partition. The designation of classes and divisions is defined in Chapter 5, Article 500, of NFPA 70, *National Electrical Code.*

NFPA 70 recognizes two divisions of hazard with respect to flammable vapors. Equipment listed for Division 1 confines all actual or potential sparking equipment within construction. A flame, originating within the equipment in vapors which have entered, cannot propagate outside the equipment to ignite a flammable atmosphere that may exist outside. Examples of equipment suitable for Division 2 locations are three-phase squirrel cage motors (single-phase motors use a sparking switch to split the phase to get starting torque) and lights in suitable enclosures, or other effective protective means. Equipment listed for Divison 1 is used, and required, where flammable vapors are often present in explosive concentration; Divison 2 is used where they are seldom present.

5-7.4 The area classifications listed in Table 5-7.3 are based on the premise that the installation meets the applicable requirements of this code in all respects. Should this not be the case, the authority having jurisdiction shall have the authority to determine the extent of the classified areas.

This warning is particularly important in connection with 5-3.5. If ventilation is inadequate, larger volumes of the atmosphere may become hazardous; also, note the last item in Table 5-7.3 having to do with the distances from drainage systems.

5-7.5 Extent of classified areas shall be as shown in Table 5-7.3.

5-7.6 Where the provisions of 5-7.1, 5-7.2, 5-7.3, 5-7.4 and 5-7.5 require the installation of electrical equipment suitable for Class I, Division 1 or Division 2 locations, ordinary electrical equipment including switchgear may be used if installed in a room or enclosure which is maintained under positive pressure with respect to the classified area. Ventilation makeup air shall be uncontaminated by flammable

---

® Registered trademark of the National Fire Protection Association, Inc.

vapors. NFPA 496, *Standard for Purged Enclosures for Electrical Equipment in Hazardous Locations*, provides details for these types of installations.

> Obviously, the make-up air should also be free from flammable gases or from toxic or corrosive materials.

5-8 Repairs to Equipment.

5-8.1 Hot work, such as welding or cutting operations, use of spark-producing power tools, and chipping operations shall be permitted only under supervision of an individual in responsible charge. The individual in responsible charge shall make an inspection of the area to be sure that it is safe for the work to be done and that safe procedures will be followed for the work specified. NFPA 327, *Standard Procedures for the Standard for Cleaning or Safeguarding Small Tanks and Containers*, and NFPA 36, *Standard for Solvent Extraction Plants*, provide information on such operations.

> Usually, the responsible person in charge issues a formal permit for hot work. It is important that this be for a specific job and that a new one, if the work is incomplete, be issued after each change in shift or other work interruption. Also, see the discussion following 8-7.2.2.

5-9 Housekeeping.

5-9.1 Maintenance and operating practices shall be in accordance with established procedures which will tend to control leakage and prevent the accidental escape of flammable or combustible liquids. Spills shall be cleaned up promptly.

> In the case of Class I liquids, methods used in cleanup should minimize the escape of flammable vapors. If the liquids are not water soluble (so that dilution would not help), vapors may be directed in some desired direction by the use of water spray. Soaking up the spill with diatomaceous earth or expanded vermiculite is also helpful. There are many reliable commercially available materials.

5-9.2 Adequate aisles shall be maintained for unobstructed movement of personnel and so that fire protection equipment can be brought to bear on any part of flammable or combustible liquid storage, use, or any unit physical operation.

5-9.3 Combustible waste material and residues in a building or unit operating area shall be kept to a minimum, stored in covered metal receptacles and disposed of daily.

5-9.4 Ground area around buildings and unit operating areas shall be kept free of weeds, trash or other unnecessary combustible materials.

Table 5-7.3

Location	NEC Class I Division	Extent of Classified Area
Indoors equipment installed in accordance with 5-3.5.2 where flammable vapor-air mixtures may exist under normal operations.	1	Area within 5 feet of any edge of such equipment, extending in all directions.
	2	Area between 5 feet and 8 feet of any edge of such equipment, extending in all directions. Also, area up to 3 feet above floor or grade level within 5 feet to 25 feet horizontally from any edge of such equipment.*
Outdoor equipment of the type covered in 5-3.5.2 where flammable vapor-air mixtures may exist under normal operations.	1	Area within 3 feet of any edge of such equipment, extending in all directions.
	2	Area between 3 feet and 8 feet of any edge of such equipment extending in all directions. Also, area up to 3 feet above floor or grade level within 3 feet to 10 feet horizontally from any edge of such equipment.
TANK — ABOVEGROUND**		
Shell, Ends, or Roof and Dike Area	2	Within 10 feet from shell, ends or roof of tank. Area inside dikes to level of top of dike.
Vent	1	Within 5 feet of open end of vent, extending in all directions.
	2	Area between 5 feet and 10 feet from open end of vent, extending in all directions.
Floating Roof	1	Area above the roof and within the shell.

*The release of Class I liquids may generate vapors to the extent that the entire building, and possibly a zone surrounding it, should be considered a Class I, Division 2 location.

**For Tanks — Underground, see Section 7-5 of Chapter 7.

Table 5-7.3— *continued*

Location	NEC Class I Division	Extent of Classified Area
DRUM AND CONTAINER FILLING		
Outdoors, or Indoors with Adequate Ventilation	1	Within 3 feet of vent and fill opening, extending in all directions.
	2	Area between 3 feet and 5 feet from vent or fill opening, extending in all directions. Also up to 18 inches above floor or grade level within a horizontal radius of 10 feet from vent or fill opening.
PUMPS, BLEEDERS, WITHDRAWAL FITTINGS, METERS AND SIMILAR DEVICES		
Indoors	2	Within 5 feet of any edge of such devices, extending in all directions. Also up to 3 feet above floor or grade level within 25 feet horizontally from any edge of such devices.
Outdoors	2	Within 3 feet of any edge of such devices, extending in all directions. Also up to 18 inches above grade level within 10 feet horizontally from any edge of such devices.
PITS		
Without Mechanical Ventilation	1	Entire area within pit if any part is within a Division 1 or 2 classified area.
With Mechanical Ventilation	2	Entire area within pit if any part is within a Division 1 or 2 classified area.
Containing Valves, Fittings or Piping, and Not Within a Division 1 or 2 Classified Area	2	Entire pit.
DRAINAGE DITCHES, SEPARATORS, IMPOUNDING BASINS	2	Area up to 18 inches above ditch, separator or basin. Also up to 18 inches above grade within 15 feet horizontally from any edge.

SI Units: 1 in. = 25.40 mm; 1 ft = 0.3048 m.

The first item in Table 5-7.3 is based on the premise that ventilation at least as good as required in 5-3.5.2 will always be functioning when the equipment is in use. For protection of employees' health, better ventilation will usually exist. The first footnote points out that the area classifications are based on normal use and that it may be desirable to consider accidental spills or failure of mechanical ventilation.

Items 2 and 3, "Outdoor Equipment" and "Aboveground Tanks," consider situations where, as discussed following 2-2.6.2, air movement, even on nominally calm days, causes rapid dilution of vapors.

Item 4, "Drum and Container Filling," does not differentiate between indoor and outdoor locations since the rate of filling and the exposed liquid surfaces are small.

Item 5, "Pumps, Bleeders, Withdrawal Fittings, Meters, and Similar Devices," distinguishes between indoor and outdoor installations because there is greater potential for larger leaks, and indoor ventilation is usually poorer than outdoor.

Item 6, "Pits," considers piping as a hazard source. This is not because leaks are anticipated in pipe within the pit, but because leaks external to the pit can find their way into it by way of loose earth in the backfill around the pipe and poorly sealed openings in the pit walls. Flammable liquid vapors also tend to collect and settle in low spots.

In Item 7, "Drainage Ditches, Separators, Impounding Basins," the 15 ft (4.57 m) distance, rather than the 10 ft (3 m) indicated elsewhere, is based on the fact that turbulence induced by flow will tend to increase vapor generation.

REFERENCES CITED BY CODE

(*These publications comprise a part of the requirements to the extent called for by the* Code.)

NFPA 13, *Standard for Installation of Sprinkler Systems*, NFPA, Boston, 1980.

NFPA 15, *Water Spray Fixed Systems*, NFPA, Boston, 1979.

NFPA 327, *Standard for Cleaning or Safeguarding Small Tanks and Containers*, NFPA, Boston, 1975.

NFPA 36, *Standard for Solvent Extraction Plants*, NFPA, Boston, 1978.

NPFA 496, *Standard for Purged and Pressurized Enclosures for Electrical Equipment*, NFPA, Boston, 1974.

NFPA 70, *National Electrical Code*, NFPA, Boston, 1981.

NFPA 72D, *Standard for Proprietary Protective Signaling Systems*, NFPA, Boston, 1979.

NFPA 78, *Lightning Protection Code*, NFPA, Boston, 1977.

NFPA 90A, *Air Conditioning and Ventilating Systems*, NFPA, Boston, 1978.

NFPA 91, *Blower and Exhaust Systems, Dust, Stock and Vapor Removal or Conveying*, NFPA, Boston, 1973.

NFPA 77, *Recommended Practice on Static Electricity*, NFPA, Boston, 1977.

REFERENCES CITED IN COMMENTARY

[1]ASMI *Boiler and Pressure Vessel Code,* American Society of Mechanical Engineers, New York, 1980.

[2]NFPA 15, *Standard for Water Spray Fixed Systems,* NFPA, Boston, 1979.

[3]National Fire Protection Association, *The Systems Approach to Fire Protection,* Boston, 1976.

[4]NFPA 70, *National Electrical Code,* NFPA, Boston, 1981.

6

Bulk Plants and Terminals

In Section 1-2, bulk plants are defined as facilities where liquids are received by connections to pipe lines or in bulk from tank cars, tank vehicles, tank vessels, or barges, and where they are stored in tanks. Some blending may take place. The liquids are then shipped out by the same means to other terminals, or to the premises of consumers. The word "terminals" was previously used to indicate that liquids are supplied primarily for the distribution of liquids to the ultimate user. This distinction has now been abandoned because the requirements for fire safety are essentially the same.

6-1 Storage.

6-1.1 Class I liquids shall be stored in closed containers, or in storage tanks aboveground outside of buildings, or underground in accordance with Chapter 2.

> Although Chapter 2, Tank Storage, permits tanks for Class I liquids inside of buildings, this prohibits such inside tanks in bulk plants or terminals.

6-1.2 Class II and Class III liquids shall be stored in containers, or in tanks within buildings or aboveground outside of buildings, or underground in accordance with Chapter 2.

6-1.3 Containers of liquids when piled one upon the other shall be separated by dunnage sufficient to provide stability and to prevent excessive stress on container walls. The height of pile shall be consistent with stability and strength of containers.

> Liquids are seldom stored in containers at the larger pipe line terminals. In the smaller bulk plants however, and particularly in rural areas, container storage becomes very important. Such storage is permitted within buildings to afford protection from the weather and from pilferage. There are no restrictions as to quantity, beyond the requirement that storage piles be stable and limited in height to avoid damaging the containers.

6-1.4 Piping, Valves and Fittings. Piping systems shall be in accordance with Chapter 3.

6-2 Buildings.

6-2.1 Exits. Rooms in which liquids are stored or handled by pumps shall have exit facilities arranged to prevent occupants being trapped in the event of fire. NFPA 101, *Life Safety Code*, provides information on the number and location of exits.

> The requirement for exit facilities to prevent occupants from being trapped is an almost universal fire protection safeguard. However, since bulk plants and terminals do not cater to the public and normally have very few employees present, the applicability of NFPA 101[1] is limited.

6-2.2 Heating. Rooms in which Class I liquids are stored or handled shall be heated only by means not constituting a source of ignition, such as steam or hot water. Rooms containing heating appliances involving sources of ignition shall be located and arranged to prevent entry of flammable vapors.

> The word "handled" in this paragraph is intended to cover operations in which liquid is exposed to the atmosphere so as to emit vapors. It is not intended to apply to liquids in piping or in closed containers.

6-2.3 Ventilation.

6-2.3.1 Ventilation shall be provided for all rooms, buildings, or enclosures in which Class I liquids are pumped or dispensed. Design of ventilation systems shall take into account the relatively high specific gravity of the vapors. Ventilation may be provided by adequate openings in outside walls at floor level unobstructed except by louvers or coarse screens. Where natural ventilation is inadequate, mechanical ventilation shall be provided. NFPA 91, *Standard for Installation of Blower and Exhaust Systems for Dust, Stock and Vapor Removal or Conveying*, provides information on the installation of mechanical exhaust systems.

> Note, however, that dissimilar materials should not be used for blades and housings, unless the combination could not create incendive sparks.

6-2.3.2 Class I liquids shall not be stored or handled within a building having a basement or pit into which flammable vapors may travel, unless such area is provided with ventilation designed to prevent the accumulation of flammable vapors therein.

> This requirement arises from several instances in which vapor from spilled liquid migrated into basement areas and was subsequently ignited, with serious results.

6-2.3.3 Containers of Class I liquids shall not be drawn from or filled within buildings unless provision is made to prevent the accumulation of flammable vapors in hazardous concentrations. Where mechanical ventilation is required, it shall be kept in operation while flammable liquids are being handled.

This recognizes that small amounts in hazardous concentration are acceptable.

6-3 Loading and Unloading Facilities.

6-3.1 Tank vehicle and tank car loading or unloading facilities shall be separated from aboveground tanks, warehouses, other plant buildings or nearest line of adjoining property that can be built upon by a distance of at least 25 ft (7.62 m) for Class I liquids and at least 15 ft (4.57 m) for Class II and Class III liquids, measured from the nearest position of any fill spout. Buildings for pumps or shelters for personnel may be a part of the facility.

This separation is required primarily to give a chance to control a fire originating at a tank vehicle before it spreads to property on the site or the property of others.

6-3.2 Equipment such as piping, pumps, and meters used for the transfer of Class I liquids between storage tanks and the fill stem of the loading rack shall not be used for the transfer of Class II or Class III liquids.

This avoids the possibility that Class II or III liquids can be contaminated with Class I liquid which would expose the ultimate consumer to an unrecognized ignition risk.

6-3.3 Top Loading.

6-3.3.1 When top loading a tank vehicle with Class I and Class II liquids without vapor control, valves used for the final control of flow shall be of the self-closing type and shall be manually held open except where automatic means are provided for shutting off the flow when the vehicle is full.

6-3.3.2 When top loading a tank vehicle with vapor control, flow control shall be in accordance with 6-3.4.1 and 6-3.4.2.

These requirements are intended to provide safeguards against the overfilling of tank vehicles which would result in a fuel spill, with possible spread and ignition of vapor. When the former practice of filling vehicles through a top opening prevailed and an operator was required to be on the top of the tank to observe the fill marker, a valve held open manually provided an acceptable safeguard. With the advent of bottom filling it became unnecessary for the operator to be on top of the tank, and additional safeguards were required. A hand-held selfclosing valve would not serve the purpose unless some completely

reliable means could be provided to tell the operator when the tank was full. The usual solution has been to employ a preset stopmeter which will automatically stop the liquid flow when the preset quantity has been delivered into the tank. This safeguard has not been deemed to be completely adequate alone, and the additional requirement of an automatic device, electrical or otherwise, which will stop the flow when the tank is full, has been required. This dual shutoff has also been considered adequate where vapor recovery is employed, so that it is unnecessary to open the top openings of the tank during filling.

6-3.4 Bottom Loading.

6-3.4.1 When bottom loading a tank vehicle, with or without vapor control, a positive means shall be provided for loading a predetermined quantity of liquid, together with an automatic secondary shut-off control to prevent overfill. The connecting components between the loading rack and the tank vehicle required to operate the secondary control shall be functionally compatible.

6-3.4.2 When bottom loading a tank vehicle that is equipped for vapor control, but when vapor control is not used, the tank shall be vented to the atmosphere to prevent pressurization of the tank. Such venting shall be at a height not lower than the top of the cargo tank on the vehicle.

The last sentence points out that vapor must be released where there is a reasonable chance for dissipation. Vapor return connections are frequently at about the same level as the liquid connections. If vapor return is not being utilized, some means must be provided to release the vapor at least as high as the top of the tank, or to pipe the vapor to a remote safe point.

6-3.4.3 When bottom loading a tank vehicle, the coupling between the liquid loading hose or pipe and the truck piping shall be by means of a dry disconnect coupling.

6-3.5 Connections to the plant vapor control system shall be designed to prevent the escape of vapor to the atmosphere when not connected to a tank vehicle.

6-3.6 Vapor processing equipment shall be separated from aboveground tanks, warehouses, other plant buildings, loading and unloading facilities, or nearest line of adjoining property that can be built upon, by a distance of at least 25 ft (7.62 m). Vapor processing equipment shall be protected against physical damage by remote location or by the provision of guard rails, curbs, or fencing.

6-3.7 Static Protection. Bonding facilities for protection against static sparks during the loading of tank vehicles through open domes shall be provided (a) where Class I liquids are loaded, or (b) where Class II or Class III liquids are loaded into vehicles which may contain vapors from previous cargoes of Class I liquids.

Filling tank vehicles through open domes at high rates has proved to be a dangerous practice for which bonding is a partial but not a complete answer. Bonding the vehicle to the fill stem can prevent the existence of a static electric potential between the fill stem and the side of the fill opening, where such a spark could ignite vapor being displaced when the tank is filled. This requirement applies both when loading Class I liquids and when loading Class II or Class III liquids into tanks which may contain vapors from a previous cargo that contained Class I liquids. In practice, bonding should be required at all racks loading liquids because the operator usually has no knowledge of the nature of the cargo previously carried. See exceptions in 6-3.7.3.

Bonding, as described in 6-3.7.1 and 6-3.7.2, cannot be relied upon to prevent an electric discharge inside a tank under the conditions described in 6-3.7.4 because the charge resides on the surface of the oil which cannot be bonded. NFPA 77[2] covers the problem in greater detail. Other precautions against possible ignition from static electricity are given in 6-3.9. A static charge will be developed by the simple flow of liquid through the fill stem. Bonding is necessary to drain off this static charge developed in the fill stem.

6-3.7.1 Protection as required in 6-3.7 shall consist of a metallic bond wire permanently electrically connected to the fill stem or to some part of the rack structure in electrical contact with the fill stem. The free end of such wire shall be provided with a clamp or equivalent device for convenient attachment to some metallic part in electrical contact with the cargo tank of the tank vehicle.

6-3.7.2 Such bonding connection shall be made fast to the vehicle or tank before dome covers are raised and shall remain in place until filling is completed and all dome covers have been closed and secured.

6-3.7.3 Bonding as specified in 6-3.7, 6-3.7.1, and 6-3.7.2 is not required:

(a) where vehicles are loaded exclusively with products not having a static accumulating tendency, such as asphalts, including cutback asphalts, most crude oils, residual oils and water soluble liquids;

(b) where no Class I liquids are handled at the loading facility and the tank vehicles loaded are used exclusively for Class II and Class III liquids; and

(c) where vehicles are loaded or unloaded through closed bottom or top connections whether the hose or pipe is conductive or nonconductive.

6-3.7.4 Filling through open domes into the tanks of tank vehicles or tank cars that contain vapor-air mixtures within the flammable range, or where the liquid being filled can form such a mixture, shall be by means of a downspout which extends near the bottom of the tank. This precaution is not required when loading liquids which are nonaccumulators of static charges. NFPA 77, *Recommended Practice on Static Electricity*, provides additional information on static electricity protection.

The bonding methods specified cover all situations where bonding can be effective in preventing the existence of a static electric potential between fill stem and fill opening, and also situations where bonding is unnecessary and not required. However, bonding cannot be totally relied upon to prevent static problems *inside* the tank because a charged liquid surface is inaccessible for bonding unless the liquid is conductive. If the liquid is nonconducting, it is difficult for a charge to dissipate in a form other than a nonincendive corona.

The precautions in 6-3.7.4 are a partial answer. NFPA 77 covers the problem in greater detail. Other precautions are given in 6-3.8 and 6-3.9.

6-3.8 Stray Currents. Tank car loading facilities where flammable and combustible liquids are loaded or unloaded through open domes shall be protected against stray currents by permanently bonding the pipe to at least one rail and to the rack structure, if of metal. Multiple pipes entering the rack area shall be permanently electrically bonded together. In addition, in areas where excessive stray currents are known to exist, all pipe entering the rack area shall be provided with insulating sections to electrically isolate the rack piping from the pipe lines. These precautions are not necessary where Class II or Class III liquids are handled exclusively and there is no probability that tank cars will contain vapors from previous cargoes of Class I liquids. Temporary bonding is not required between the tank car and the rack or piping during either loading or unloading irrespective of the class of liquid handled.

Stray currents at a tank car loading facility can result if there is some major defect in an electrical installation in the area. If it can be discovered, the cause should be identified and corrected. Permanent bonds between the rack piping and the track are an unsatisfactory solution to the problem. If stray currents persist and cannot be controlled, the best practice is to insulate the piping at the rack from its supply lines and bond the rack piping to one rail adjacent to the rack. While precautions would not, theoretically, be necessary where only Class III liquids are handled, there is no assurance that a car could not contain flammable vapors from a prior cargo. Temporary bonds applied at the time of loading or unloading are likely to be inadequate and should not be required. Note that stray currents are not considered a hazard in the case of tank vehicles because of the insulation afforded by the rubber tires.

6-3.9 Container Filling Facilities. Class I liquids shall not be dispensed into metal containers unless the nozzle or fill pipe is in electrical contact with the container. This can be accomplished by maintaining metallic contact during filling, by a bond wire between them, or by other conductive path having an electrical resistance not greater than 10^6 ohms. Bonding is not required where a container is filled through a closed system, or is made of glass or other non-conducting material. NFPA 77, *Recommended Practice on Static Electricity*, provides information on static protection.

The purpose of these requirements in to ensure that there will not be a difference of potential between a metallic nozzle or a fill pipe and a metal container during the dispensing process. Such a potential could be generated by the flow of the fluid through the nozzle. The restriction is limited to Class I liquids because other liquids, unless heated, would not produce an ignitible vapor-air mixture at the opening. Potential difference is prevented by metallic contact or by the use of a ground wire. In some instances metallic contact is inherent in the installation, as when steel drums are filled while resting on a metallic floor or conveyor which is at the same potential as the nozzle or fill pipe.

6-4 Wharves.

6-4.1 The term wharf shall mean any wharf, pier, bulkhead or other structure over or contiguous to navigable water, the primary function of which is the transfer of liquid cargo in bulk between shore installations and any tank vessel, ship, barge, lighter boat or other mobile floating craft; and this section shall apply to all such installations except marine service stations as covered in Chapter 7. If liquids are handled in bulk quantities across general purpose piers or wharves, NFPA 87, *Standard for the Construction and Protection of Piers and Wharves*, shall be followed.

This definition is broad enough to cover structures ranging in size from a small finger pier or bulkhead to a structure intended for loading or unloading large tankers, the principal limitation being that the primary function is the transfer of liquid cargo. The various precautions listed in this section are intended to minimize the chance of escape of liquid that would float on water and which would have serious consequences if ignited on the water.

6-4.1.1 Package cargo of liquids, including full and empty drums, bulk fuel and stores may be handled over a wharf during cargo transfer at such times and places as agreed upon by the wharf superintendent and the senior deck officer on duty.

The handling of package cargo and stores during cargo transfer does not constitute an unreasonable risk if the activity is confined to areas remote from the cargo loading connection. It is the joint responsibility of the ships' officers and the wharf representatives to prescribe times and places where such transfer can be safely carried out.

6-4.1.2 Wharves at which liquid cargoes are to be transferred in bulk quantities to or from tank vessels shall be at least 100 ft (30.48 m) from any bridge over a navigable waterway, or from an entrance to or superstructure of any vehicular or railroad tunnel under a waterway. The termination of the wharf loading or unloading fixed piping shall be at least 200 ft (60.96 m) from a bridge or from an entrance to or superstructure of a tunnel.

This description is obviously intended to apply to wharves where large quantities of liquid are handled, and would not ordinarily be applied to a small barge loading facility.

6-4.2 Substructure and deck shall be substantially designed for the use intended. Deck may employ any material which will afford the desired combination of flexibility, resistance to shock, durability, strength and fire resistance. Heavy timber construction is acceptable.

Because of the wide variety in size and use of wharves, it is impossible to prescribe details of construction. It becomes an engineering problem to determine which type of material or manner of construction will best meet the requirements for the particular operation intended.

6-4.3 Tanks used exclusively for ballast water or Class II or Class III liquids may be installed on suitably designed wharves.

The installation of tanks for Class II or III liquids is permitted on structurally adequate wharves because of the reduced fire risk with these liquids. Ballast water tanks can be installed on the wharves for the same reason, although it is obvious that the problem of supporting large tanks will usually dictate that they be installed on shore.

6-4.4 Loading pumps capable of building up pressures in excess of the safe working pressure of cargo hose or loading arms shall be provided with by-passes, relief valves, or other arrangement to protect the loading facilities against excessive pressure. Relief devices shall be tested at not more than yearly intervals to determine that they function satisfactorily at the pressure at which they are set.

6-4.4.1 All pressure hoses and couplings shall be inspected at intervals appropriate to the service. With the hose extended, test the hose and couplings using the "in service maximum operating pressures." Any hose showing material deteriorations, signs of leakage, or weakness in its carcass or at the couplings shall be withdrawn from service and repaired or discarded.

6-4.5 Piping, valves and fittings shall be in accordance with Chapter 3, with the following exceptions and additions:

6-4.5.1 Flexibility of piping shall be assured by appropriate layout and arrangement of piping supports so that motion of the wharf structure resulting from wave action, currents, tides or the mooring of vessels will not subject the pipe to repeated strain beyond the elastic limit.

Assuring adequate flexibility of piping on wharves is important because a wharf structure is usually flexible and subject to movement from the action of wind, waves, and possible impact of vessels during mooring. Failure of piping under these circumstances can release liquid which can, if ignited, result in a serious fire. Spread of burning liquid on the water is likely to endanger all of the wharf piping, since much of the piping is frequently installed beneath wharf

surface to provide clear space for handling of dry cargo. In addition, the wharf structure will not be able to withstand the effects of a protracted fire in most instances.

6-4.5.2 Pipe joints depending upon the friction characteristics of combustible materials or grooving of pipe ends for mechanical continuity of piping shall not be used.

Pipe joints of the preceding types have been found to be particularly susceptible to damage from short-term fire exposure, releasing contents which intensify the fire.

6-4.5.3 Swivel joints may be used in piping to which hoses are connected, and for articulated swivel-joint transfer systems, provided that the design is such that the mechanical strength of the joint will not be impaired if the packing material should fail, as by exposure to fire.

This requirement is made to ensure that swivel-joint transfer systems do not become completely disengaged under fire exposure.

6-4.5.4 In addition to the requirements of 3-6.1, each line conveying Class I and Class II liquids leading to a wharf shall be provided with a readily accessible block valve located on shore near the approach to the wharf and outside of any diked area. Where more than one line is involved, the valves shall be grouped in one location.

The intent here is that pipe lines that lead to a wharf be equipped with block valves located on shore, so as to be accessible for operation even if fire in liquid that is floating on the water makes it impossible to reach valves located on the wharf itself. In some places it may be desirable to install additional block valves on wharf lines at points where a group of loading spots branch off from the main lines, as on a "T" headed wharf.

6-4.5.5 Means of easy access shall be provided for cargo line valves located below the wharf deck.

The intent is to make it easy to operate valves located below the wharf deck, by means of trap doors or the like, even though it is necessary to keep the wharf surface unobstructed for the passage of motor vehicles. Such trap doors are usually conspicuously painted, and the parking of vehicles or storage of dry cargo over them is forbidden.

6-4.5.6 Pipe lines on wharves shall be adequately bonded and grounded if Class I and Class II liquids are handled. If excessive stray currents are encountered, insulating joints shall be installed. Bonding and grounding connections on all pipe lines shall be located on wharf side of hose riser insulating flanges, if used, and shall be accessible for inspection.

> Wharf pipe lines usually have welded flange or screwed connections and are thus electrically continuous, eliminating the need for any special bonding. However, if for any reason connections which are not conducting are installed, a survey should be made to determine the desirability of placing electrical bonds around them. Insulating connections provided at connections for cargo hose are a deliberate attempt to isolate the vessel from possible stray currents derived from shoreside installations, and bonds around them are obviously undesirable. If the presence of continuous or intermittent stray currents in shore piping is revealed by test, all wharf piping should be electrically isolated from the shore piping with insulating connections. In some cases it will be desirable to provide a deliberate ground connection for wharf pipe lines. See NFPA 77.

6-4.5.7 Hose or articulated swivel-joint pipe connections used for cargo transfer shall be capable of accommodating the combined effects of change in draft and maximum tidal range, and mooring lines shall be kept adjusted to prevent surge of the vessel from placing stress on the cargo transfer system.

> This is an important operating requirement and is the responsibility of both the vessel and wharf personnel.

6-4.5.8 Hose shall be supported so as to avoid kinking and damage from chafing.

6-4.6 Suitable portable fire extinguishers with a rating of not less than 20-B shall be located within 75 ft (22.8 m) of those portions of the facility where fires are likely to occur, such as hose connections, pumps and separator tanks.

6-4.6.1 Where piped water is available, ready-connected fire hose in size appropriate for the water supply shall be provided so that manifolds where connections are made and broken can be reached by at least one hose stream.

> While there is probably little that can be accomplished with a water stream in case of failure of cargo hose or other connections in the manifold area, it seems prudent that where water is available at whatever pressure required for ship's services or other uses, appropriate fire hose be provided to make use of it.

6-4.6.2 Material shall not be placed on wharves in such a manner as to obstruct access to fire fighting equipment or important pipe line control valves.

6-4.6.3 Where the wharf is accessible to vehicle traffic, an unobstructed roadway to the shore end of the wharf shall be maintained for access of fire fighting apparatus.

6-4.7 Loading or discharging shall not commence until wharf superintendent and officer in charge of tank vessel agree that tank vessel is properly moored and all connections are properly made.

6-4.7.1 Mechanical work shall not be performed on the wharf during cargo transfer, except under special authorization based on a review of the area involved, methods to be employed, and precautions necessary.

6-5 Electrical Equipment.

6-5.1 This section shall apply to areas where Class I liquids are stored or handled. For areas where Class II or Class III liquids only are stored or handled, the electrical equipment may be installed in accordance with the provisions of NFPA 70, *National Electrical Code*, for ordinary locations.

6-5.2 All electrical equipment and wiring shall be of a type specified by and shall be installed in accordance with NFPA 70, *National Electrical Code*.

6-5.3 So far as it applies, Table 6-5.3 shall be used to delineate and classify areas for the purpose of installation of electrical equipment under normal circumstances. In Table 6-5.3 a classified area shall not extend beyond an unpierced wall, roof or other solid partition. The designation of classes and divisions is defined in Chapter 5, Article 500, of NFPA 70, *National Electrical Code*.

See Figure 6-1.

Table 6-5.3, which is largely self-explanatory, provides a guide for assigning electrical classifications in all areas in bulk plants. It must be kept in mind that frequent changes in classification between closely spaced areas are undesirable, and it is preferable to avoid unnecessary changes in classification which would result in an increased number of seals when conduit is used for the wiring.

6-5.4 The area classifications listed in 6-5.3 shall be based on the premise that the installation meets the applicable requirements of this code in all respects. Should this not be the case, the authority having jurisdiction shall have the authority to classify the extent of the classified area.

6-6 Sources of Ignition.

6-6.1 Class I liquids shall not be handled, drawn, or dispensed where flammable vapors may reach a source of ignition. Smoking shall be prohibited except in designated localities. "NO SMOKING" signs shall be conspicuously posted where hazard from flammable vapors is normally present; NFPA 78, *Lightning Protection Code*, provides information on lightning protection.

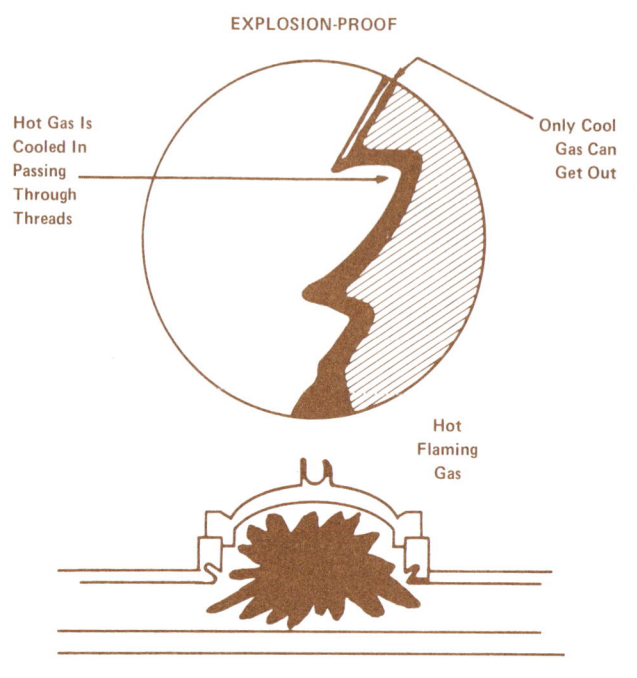

166 FLAMMABLE AND COMBUSTIBLE LIQUIDS CODE HANDBOOK

Figure 6-1. Although electrical equipment installation in Division 2 locations is considerably less restrictive than for Division 1, the basic design criterion for all Class I locations is "explosionproof." All equipment must withstand internal explosions of flammable gas or vapor-air mixtures. It is recognized that surrounding flammable gas or vapor-air mixtures will, under certain conditions, enter the enclosure of the equipment, creating the possibility of ignition of these mixtures within the enclosure. The use of explosionproof equipment, or equipment that is either totally sealed or enclosed within an inert gas-filled environment, is designed to prevent the propagation of flame to the outside surrounding atmosphere, which may also contain flammable vapor-air mixtures.

6-7 Drainage and Waste Disposal.

6-7.1 Provision shall be made to prevent liquids which can be spilled at loading or unloading points from entering public sewers and drainage systems, or natural waterways. Connection to such sewers, drains, or waterways by which liquids might enter shall be provided with separator boxes or other approved means whereby such entry is precluded. Crankcase drainings and liquids shall not be dumped into sewers, but shall be stored in tanks or tight drums outside of any building until removed from the premises.

6-8 Fire Control.

6-8.1 Listed portable fire extinguishers of appropriate size, type and number shall be provided. NFPA 10, *Standard for Portable Fire Extinguishers*, provides information on this subject. At least one extinguisher with a minimum classification of 20-B shall be provided at each loading or unloading facility. Where piped

Table 6-5.3 — Electrical Equipment Classified Areas — Bulk Plants

Location	NEC Class I, Group D Division	Extent of Classified Area
TANK VEHICLE AND TANK CAR* Loading Through Open Dome	1	Within 3 feet of edge of dome, extending in all directions.
	2	Area between 3 feet and 15 feet from edge of dome, extending in all directions.
Loading Through Bottom Connections With Atmospheric Venting	1	Within 3 feet of point of venting to atmosphere extending in all directions.
	2	Area between 3 feet and 15 feet from point of venting to atmosphere, extending in all directions. Also up to 18 inches above grade within a horizontal radius of 10 feet from point of loading connection.
Loading Through Closed Dome With Atmospheric Venting	1	Within 3 feet of open end of vent, extending in all directions.
	2	Area between 3 feet and 15 feet from open end of vent, extending in all directions. Also within 3 feet of edge of dome, extending in all directions.
Loading Through Closed Dome With Vapor Control	2	Within 3 feet of point of connection of both fill and vapor lines, extending in all directions.
Bottom Loading With Vapor Control. Any Bottom Unloading	2	Within 3 feet of point of connections extending in all directions. Also up to 18 inches above grade within a horizontal radius of 10 feet from point of connection.

*When classifying extent of area, consideration shall be given to fact that tank cars or tank vehicles may be spotted at varying points. Therefore, the extremities of the loading or unloading positions shall be used.

Table 6-5.3 — *continued*

Location	NEC Class I, Group D Division	Extent of Classified Area
DRUM AND CONTAINER FILLING		
Outdoors, or Indoors With Adequate Ventilation	1	Within 3 feet of vent and fill opening, extending in all directions.
	2	Area between 3 feet and 5 feet from vent or fill opening, extending in all directions. Also up to 18 inches above floor or grade level within a horizontal radius of 10 feet from vent or fill opening.
TANK — ABOVEGROUND*		
Shell, Ends, or Roof and Dike Area	2	Within 10 feet from shell, ends, or roof of tank. Area inside dikes to level of top of dike.
Vent	1	Within 5 feet of open end of vent, extending in all directions.
	2	Area between 5 feet and 10 feet from open end of vent, extending in all directions.
Floating Roof	1	Area above the roof and within the shell.
PITS		
Without Mechanical Ventilation	1	Entire area within pit if any part is within a Division 1 or 2 classified area.
With Mechanical Ventilation	2	Entire area within pit if any part is within a Division 1 or 2 classified area.
Containing Valves, Fittings or Piping, and Not Within a Division 1 or 2 Classified Area	2	Entire pit.

*For Tanks — Underground, see Section 7-5 of Chapter 7.

Table 6-5.3 — *continued*

Location	NEC Class I, Group D Division	Extent of Classified Area
PUMPS, BLEEDERS, WITHDRAWAL FITTINGS, METERS AND SIMILAR DEVICES		
Indoors	2	Within 5 feet of any edge of such devices, extending in all directions. Also up to 3 feet above floor or grade level within 25 feet horizontally from any edge of such devices.
Outdoors	2	Within 3 feet of any edge of such devices, extending in all directions. Also up to 18 inches above grade level within 10 feet horizontally from any edge of such devices.
STORAGE AND REPAIR GARAGE FOR TANK VEHICLES	1	All pits or spaces below floor level.
	2	Area up to 18 inches above floor or grade level for entire storage or repair garage.
DRAINAGE DITCHES, SEPARATORS, IMPOUNDING BASINS	2	Area up to 18 inches above ditch, separator or basin. Also up to 18 inches above grade within 15 feet horizontally from any edge.
GARAGES FOR OTHER THAN TANK VEHICLES	Ordinary	If there is any opening to these rooms within the extent of an outdoor classified area, the entire room shall be classified the same as the area classification at the point of the opening.
OUTDOOR DRUM STORAGE	Ordinary	
INDOOR WAREHOUSING WHERE THERE IS NO FLAMMABLE LIQUID TRANSFER	Ordinary	If there is any opening to these rooms within the extent of an indoor classified area, the room shall be classified the same as if the wall, curb or partition did not exist.
OFFICE AND REST ROOMS	Ordinary	

SI Units: 1 in. = 25.40 mm; 1 ft = 0.3048 m.

water is available, ready-connected hose in size appropriate for the water supply shall be provided at locations where fires are likely to occur.

> The decision as to what constitutes "suitable fire control devices" depends on such a variety of conditions that no specific recommendations can be made. Many wharves in this category consist of little more than a wooden walkway supporting a single pipe line, with minimal mooring facilities at the end. Under such circumstances, if fire occurs, the vessel can disconnect and move away, and the investment at risk in the wharf itself is so small as to not justify elaborate fire control measures. Accessible block valves in the liquid pipe lines on shore — the one important safeguard that should not be overlooked — will limit the extent of the fire. For more elaborate facilities, possibly including buildings for the storage of ships' supplies, a more extensive fire control system may be justified.
>
> Experience has shown that fires at bulk plants are most likely to occur in connection with tank truck filling, and in many cases the fire can be so large as to preclude any effective use of first-aid fire fighting equipment. Pending arrival of public fire departments with their large equipment, much can be accomplished by closing tank valves so that continuing flow of fuel is prevented. Portable extinguishers can be useful in extinguishing fires at the fill opening of tank vehicles, although the practice of filling through open domes is decreasing. Where an elevated loading rack is provided, extinguishers should be at ground level at the stairway giving access to the rack. Fires in offices and warehouses are usually small at the start; if promptly discovered, they can be controlled with small diameter hose such as a garden hose.

6-8.2 All plant fire protection facilities shall be adequately maintained and periodically inspected and tested to make sure they are always in satisfactory operating condition, and they will serve their purpose in time of emergency.

6-8.3 Bulk plants and terminals shall have a written emergency procedure plan. The plan shall be designed to minimize the hazard to the public and to plant employees in the event of a fire or other emergency conditions. The plan shall be posted, or located in a strategic and accessible location. Plant personnel assigned to emergency duties shall be trained in these duties.

REFERENCES CITED BY *CODE*

(These publications comprise a part of the requirements to the extent called for by the **Code**.*)*

NFPA 70, *National Electrical Code*, NFPA, Boston, 1981.
NFPA 78, *Lightning Protection Code*, NFPA, Boston, 1977.
NFPA 87, *Construction & Protection, Piers and Wharves*, NFPA, Boston, 1975.

NFPA 91, *Blower and Exhaust Systems, Dust, Stock & Vapor Removal or Conveying*, NFPA, Boston, 1973.

NFPA 101, *Life Safety Code*, NFPA, Boston, 1981.

NFPA 77, *Recommended Practice on Static Electricity*, NFPA, Boston, 1977.

References Cited in Commentary

[1] NFPA 101, *Life Safety Code*, NFPA, Boston, 1981.

[2] NFPA 77, *Recommended Practice on Static Electricity*, NFPA, Boston, 1977.

7

Service Stations

The *Code* provisions for service stations differ substantially from those of other specific occupancies. The main reason for this is that dispensing or transferring flammable and combustible liquids is an integral part of their operation. As previously mentioned, dispensing or transferring operations are the most hazardous of all ordinary operations dealing with flammable and combustible liquids. Environmental protection regulations have also introduced a number of additional requirements relative to this type of occupancy, such as the provisions for vapor recovery and vapor processing systems. Public access to these occupancies is another inherent feature which causes concern relative to the fire hazard potential.

7-1 Scope.

7-1.1 This chapter applies to automotive and marine service stations and to service stations located inside buildings. Reference shall also be made to NFPA 302, *Fire Protection Standard for Motor Craft*, for safety precautions while fueling at marine service stations, and to NFPA 303, *Fire Protection Standard for Marinas and Boatyards*, for additional requirements applicable to marine service stations.

One normally thinks of service stations as public facilities. However, this chapter of the *Code* also applies to automotive service stations that are private, such as those at garages, police and fire stations, industrial plants, automotive truck terminals, and private marine service stations. (See in Section 1-2 definitions of automotive service stations, marine service stations, and service stations located inside buildings.)

7-2 Storage.

7-2.1 General Provisions.

This *Code* sets forth several general storage provisions for service stations. They deal with venting, stations adjoining bulk plants, storage and pits in basement areas, inventory records, tank storage, tanks inside buildings, and container storage on service station properties.

7-2.1.1 Liquids shall be stored in approved closed containers not exceeding 60 gal (227.1 L) capacity, in tanks in special enclosures as described in 7-2.2, in aboveground tanks as provided for in 7-2.1.5, or in tanks located underground as in Section 2-3. Vent pipes on tanks storing gasoline shall be in accordance with 2-3.5.1, 2-3.5.2 and 2-3.5.6, as applicable, and shall discharge only upward in order to disperse vapors. (*Also see 7-8.3.4 and 7-8.3.5*.)

> Tanks at service stations are generally required to be installed underground. For exceptions, see 7-2.2 and 7-2.1.5. Also see the comments on venting in 2-2.6.2 and 2-3.5.1.
>
> The requirement for tanks at service stations to discharge vapors only upward from vents prohibits the use of a U-bend or weather hood to keep out rain and foreign matter. Experience has shown that the risk of serious contamination of the tank content which accompanies upward discharge of vapor from an open vent is small. On the other hand, there have been several serious incidents of fire when vapors are directed downward. Existing service station vent pipes that have a weather hood, U-bend, or any type of vent cap that forces the vapor discharge downward, or otherwise hinders the discharge of vapors, should be corrected in order to avoid a fire problem.

7-2.1.2 Aboveground tanks, located at a bulk plant, shall not be connected by piping to service station underground tanks. Apparatus dispensing Class I liquids into the fuel tanks of motor vehicles of the public shall not be located at a bulk plant unless separated by a fence or similar barrier from the area in which bulk operations are conducted.

> This change in the 1981 edition of the *Code*, forbidding the connecting of bulk storage tanks, was deemed necessary to reduce the likelihood of overfilling underground tanks where buılk storage might result in excessively large spills.

7-2.1.3 Class I liquids shall not be stored or handled within a building having a basement or pit into which flammable vapors can travel, unless such area is provided with ventilation which will prevent the accumulation of flammable vapors therein.

> Again, the reader is reminded that flammable liquid vapors are heavier than air and will normally seek low levels.

7-2.1.4 Accurate daily inventory records shall be maintained and reconciled on all Class I liquid and diesel fuel storage tanks for indication of possible leakage from tanks or piping. The records shall be kept at the premises, available for inspection by the enforcing authority, and shall include, as a minimum, records showing by product, daily reconciliation between sales, use, receipts, and inventory on hand. If there is more than one system consisting of a tank(s) serving separate pump(s) or dispenser(s) for any product, the reconciliation shall be maintained separately for each tank system. API Publication 1621, *Recommended*

Practice for Bulk Liquid Stock Control at Retail Outlets, provides information on this subject.*

> Numerous fires and explosions have resulted from leaking underground service station tanks. In some cases, nearby buildings have been evacuated, sometimes for several weeks, because of the presence of flammable vapors. Frequently, the source of the leak is difficult to locate. Gasoline from a leaking underground tank has been known to spread in the water table for distances up to three miles. Because of this leakage problem, the *Code* specifies the need for maintaining accurate daily inventory records. If such records are maintained and discrepancies reconciled, leaks can often be discovered before they constitute a hazard.
>
> NFPA 329[1] provides guidance in determining the source of a leak, as well as procedures for testing tanks and removing liquids that are trapped underground. See also API Publication 1621, *Recommended Practice for Bulk Liquid Stock Control at Retail Outlets*.[2]

7-2.1.5 Tanks supplying marine service stations and pumps not integral with the dispensing unit shall be on shore or on a pier of the solid-fill type, except as provided in (a) and (b).

(a) Where shore location would require excessively long supply lines to dispensers, the authority having jurisdiction may authorize the installation of tanks on a pier provided that applicable portions of Chapter 2 relative to spacing, diking and piping are complied with and the quantity so stored does not exceed 1,100 gal (4163.5 L) aggregate capacity.

(b) Shore tanks supplying marine service stations may be located aboveground where rock ledges or high water tables make underground tanks impractical.

> A marine station must be located where there is adequate water depth to accommodate the vessels served. Suitable locations are frequently near steep or rocky shores where excavation to completely bury a tank would be difficult. A tank should preferably be above the high water level to avoid water in the excavation, possible corrosion, and leakage. This paragraph and the following, 7-2.1.6, recognize these risks and provide means for avoiding them.

7-2.1.6 Where tanks are at an elevation which produces a gravity head on the dispensing unit, the tank outlet shall be equipped with a device, such as a solenoid valve, positioned adjacent to and downstream from the valve specified in 2-2.7.1, so installed and adjusted that liquid cannot flow by gravity from the tank in case of piping or hose failure when the dispenser is not in use.

7-2.2 Special Enclosures.

7-2.2.1 When installation of tanks in accordance with Section 2-3 is impractical

---
*Available from American Petroleum Institute, 2101 L St. N.W., Washington, DC 20037.

because of property or building limitations, tanks for liquids may be installed in buildings if enclosed as described in 7-2.2.2 and upon specific approval of the authority having jurisdiction.

7-2.2.2 Enclosure shall be substantially liquid- and vaportight without backfill. Sides, top and bottom of the enclosure shall be of reinforced concrete at least 6 in. (152.4 mm) thick, with openings for inspection through the top only. Tank connections shall be so piped or closed that neither vapors nor liquid can escape into the enclosed space. Means shall be provided to use portable equipment to discharge to the outside any liquid or vapors which might accumulate should leakage occur.

> The object is to ensure that overfilling of the tank cannot release liquid into adjacent spaces, and to provide access to the space between the tank and the enclosure to facilitate ventilation and the removal of liquid.

7-2.2.3 At automotive service stations provided in connection with tenant or customer parking facilities in large buildings of commercial, mercantile or residential occupancy, tanks containing Class I liquids installed in accordance with 7-2.2.2 shall not exceed 6,000 gal (22 710 L) individual or 18,000 gal (68 130 L) aggregate capacity.

7-2.3 Inside Buildings.

7-2.3.1 Except where stored in tanks as provided in 7-2.2, no Class I liquids shall be stored within any service station building except in closed containers of aggregate capacity not exceeding 120 gal (454.2 L). One container not exceeding 60 gal (227.1 L) capacity equipped with a listed pump is permitted.

> The container equipped with a listed pump referred to in the last sentence should be included within that maximum capacity of 120 gal (454.2 L). The need for such a container and pump was much greater when alcohol was the predominant antifreeze or when white gasoline was dispensed for gasoline stoves and lanterns. The purpose behind permitting just one container to be equipped with a pump was to allow a second such container for a back-up supply.

7-2.3.2 Class I liquids may be transferred from one container to another in lubrication or service rooms of a service station building provided the electrical installation complies with Table 7-5 and provided that any heating equipment complies with Section 7-7. See also 7-9.1 for other possible sources of ignition.

7-2.3.3 Class II and Class III liquids may be stored and dispensed inside service station buildings from tanks of not more than 120 gal (454.2 L) each.

> The 120 gal (454.2 L) limitation allows items such as kerosene or lube oils to be dispensed from the typical rectangular tanks used at service stations.

7-3 Piping, Valves and Fittings.

7-3.1 The design, fabrication, assembly, test and inspection of the piping system shall be in accordance with Chapter 3 except that, where dispensing is from a floating structure, suitable lengths of oil-resistant flexible hose may be employed between the shore piping and the piping on the floating structure as made necessary by change in water level or shore line.

> Requirements for the design, fabrication, assembly, testing, and inspection of piping systems at service stations are basically those of Chapter 3. Major differences are the exceptions to the requirements which are included basically for marine service stations as specified in 7-3.1 and the three following paragraphs, plus the additional testing required for piping systems at service stations as specified in 7-3.1.4.

7-3.1.1 Where excessive stray currents are encountered, piping handling Class I and Class II liquids at marine service stations shall be electrically insulated from the shore piping.

> This is because shore-based stray currents might be transmitted through a conductive hose to the tank fill opening, which is often grounded to the vessel hull, making a spark possible at the fill opening.

7-3.1.2 Piping shall be located so as to be protected from physical damage.

7-3.1.3 A readily accessible valve to shut off the supply from shore shall be provided in each pipeline at or near the approach to the pier and at the shore end of each marine pipeline adjacent to the point where a flexible hose is attached.

7-3.1.4 After completion of the installation, including any paving, that section of the pressure piping system between the pump discharge and the connection for the dispensing facility shall be tested for at least 30 minutes at the maximum operating pressure of the system.

7-3.2 Remote Pumping Systems.

> At many of the newer and larger service stations, remote pumping systems are used where the pump at the tank will supply several dispensers for the same grade of gasoline. The supply lines of these systems operate under pressure. Most commercially available remote delivery pumps are installed within the tank or in a separate enclosure. See 7-3.2.3 for the exceptions.

7-3.2.1 This section shall apply to systems for dispensing Class I liquids where such liquids are transferred from storage to individual or multiple dispensing units by pumps located elsewhere than at the dispensing units.

7-3.2.2 Pumps shall be designed or equipped so that no part of the system will be subjected to pressures above its allowable working pressure. Each pump shall have installed on the discharge side an approved leak detection device which will provide an indication if the piping and dispensers are not essentially liquidtight.

> Without such a device, a leak in the piping between the pump and the dispenser might continue undetected.

7-3.2.3 Pumps installed above grade, outside of buildings, shall be located not less than 10 ft (3.05 m) from lines of adjoining property which can be built upon, and not less than 5 ft (1.5 m) from any building opening. When an outside pump location is impractical, pumps may be installed inside buildings as provided for dispensers in 7-4.1.2, or in pits as provided in 7-3.2.4. Pumps shall be substantially anchored and protected against physical damage.

7-3.2.4 Pits for subsurface pumps or piping manifolds of submersible pumps shall withstand the external forces to which they can be subjected without damage to the pump, tank or piping. The pit shall be no larger than necessary for inspection and maintenance, and shall be provided with a fitted cover.

7-3.2.5 A listed rigidly anchored emergency shutoff valve, incorporating a fusible link or other thermally actuated device, designed to close automatically in event of severe impact or fire exposure shall be properly installed in the supply line at the base of each individual island-type dispenser or at the inlet of each overhead dispensing unit. If a coupling incorporating a slip-joint feature is used to join the emergency valve to the dispenser piping, the emergency valve shall automatically close before the slip joint can disengage. The automatic closing feature of this valve shall be checked at the time of initial installation and at least once a year thereafter by manually tripping the hold-open linkage.

7-3.2.6 A vapor return pipe inside the dispenser housing shall have a shear section or flexible connector so that liquid emergency shut-off valve will function as described in 7-3.2.5.

> This valve required in 7-3.2.5 must stop the flow of gasoline in case the dispenser is physically damaged by a vehicle, or pulled over by a vehicle driving away with the nozzle still in the tank. Such a stress should release a spring-loaded mechanism which, in turn, closes the valve. Further protection for this valve is provided by a shear section which will break upon impact or severe stress. The shear section of the valve should be installed within ½ in. of the pump island level, and the entire assembly should be rigidly anchored to the island. This is to help ensure that the piping will break at the shear section, and that the spring and thermally actuated device will function to close the valve. (See Figure 7-1.)

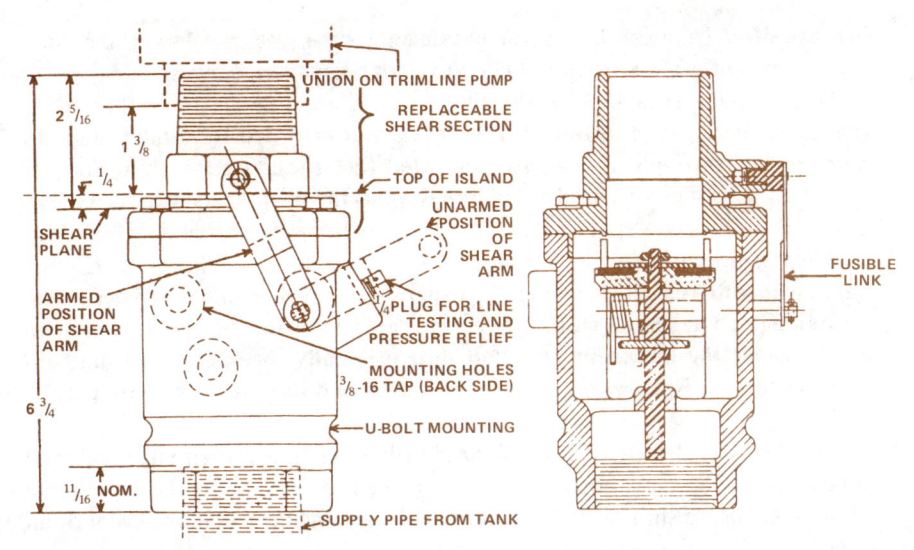

SI Units: 1 in. = 25.4 mm.

Figure 7-1. Drawings show where fusible link and shear valve are located. To trip the valve manually for the annual inspection, disconnect the fusible link from its top connection. Resetting the fusible link rearms and opens the valve. (Drawings courtesy of Gilbarco, Inc., Greensboro, NC)

7-4 Fuel Dispensing System.

7-4.1 Location of Dispensing Devices and Emergency Power Cutoff.

7-4.1.1 Dispensing devices at an automotive service station shall be so located that all parts of the vehicle being served will be on the premises of the service station. Openings beneath enclosures shall be sealed to prevent the flow of leaking fuel to lower building spaces. Dispensing devices at marine service stations may be located on open piers, wharves, floating docks, or on shore, or on piers of the solid fill type, and shall be located apart from other structures so as to provide room for safe ingress and egress of craft to be fueled. Dispensing units shall be in all cases at least 20 ft (6.1 m) from any activity involving fixed sources of ignition. Dispensing devices located inside buildings shall comply with Section 7-5.

This paragraph prohibits the installation of dispensing units on the sidewalk in front of a garage since all parts of the vehicle being served must be on the premises of the service station. The section was added to the *Code* in order to control the sources of ignition, and to decrease the possiblity of accidents to the dispenser or to the vehicle being fueled.

Sealing the openings beneath the dispenser enclosure is particularly critical for dispensing units located in parking garages or other buildings that have floor levels below the dispensing devices. This section was recently added to the *Code* after the emergency shutoff system on a dispenser at the Reversing Falls Garage in St. John, New Brunswick, failed to prevent an explosion on January 4, 1974. The emergency shutoff valve may not have functioned properly because of mechanical damage or because the pipe system below it may have been damaged when a truck knocked over the dispenser.

At the usual service station, the pipe is buried and thus firmly supported. In this case, there was no support other than that afforded by some beams the pipes passed through; the beams supported the service station floor above a basement. Gasoline was discharged into the basement where the vapors were ignited, probably by a sparking thermostatic switch that started an oil burner. The explosion wrecked the station and killed five people. The trucking company, whose driver knocked over the dispenser, was held liable for 25 percent of the damages. The service station owner was held liable for the remaining 75 percent because the basement area had not, originally, been enclosed, and the opening to the dispenser base was not sealed when the enclosure took place.

7-4.1.2 A clearly identified and easily accessible switch(es) or circuit breaker(s) shall be provided at a location remote from dispensing devices, including remote pumping systems, to shut off the power to all dispensing devices in the event of an emergency.

Since a fire or large gasoline spill at the dispensing island may make it impossible to operate the switches on the dispensing island that shut off the flow of gasoline, the *Code* requires a clearly identified emergency power cutoff to be provided at a location remote from the dispensing device. The term "clearly identified" means that a sign is to be posted indicating where the cutoff switch is located. This emergency power cutoff should be readily accessible and not blocked by storage of such things as tires or cases of lubricating oil. All service station operators as well as responding fire fighters should know the location of the emergency power cutoff.

7-4.3 Fuel Dispensing Units.

7-4.3.1 Class I liquids shall be transferred from tanks by means of fixed pumps designed and equipped to allow control of the flow and prevent leakage or accidental discharge.

7-4.3.2 Dispensing devices for Class I liquids shall be listed.

7-4.3.3 A control shall be provided that will permit the pump to operate only when a dispensing nozzle is removed from its bracket or normal position with respect to the dispensing unit, and the switch on this dispensing unit is manually actuated. This control shall also stop the pump when all nozzles have been returned, either to their brackets or to the normal nondispensing position.

The term "all nozzles" should be interpreted to mean all nozzles controlling the discharge from the pump. A proposal was submitted to the Technical Committee to allow a single-action dispensing nozzle which activates the pump upon removal from its bracket. After due consideration, the proposal was rejected.

The Committee felt that the second positive action by the user ensured greater attention to the dispensing operation. The single-action dispenser nozzle failed to provide a positive pump shutoff. The Committee felt that the proponent did not demonstrate that safety was enhanced by the single-action nozzle.

7-4.3.4 Liquids shall not be dispensed by applying pressure to drums, barrels and similar containers. Listed pumps taking suction through the top of the container or listed self-closing faucets shall be used.

See the comments following 5-2.4.5 regarding pressurizing of containers. Gravity discharge of the liquid is not permitted.

7-4.3.5 The dispensing unit and its piping, except those attached to containers, shall be mounted on a concrete island or protected against collision damage by suitable means. If located indoors, the dispenser shall also be mounted either on a concrete island or protected against collision damage by suitable means and shall be located in a position where it cannot be struck by a vehicle that is out of control descending a ramp or other slope.

7-4.4 Vapor Recovery Systems.

The United States Environmental Protection Agency, as well as state and local air pollution authorities, may soon require the recovery of gasoline vapors at automotive service stations. At the present time, laws regarding the recovery of gasoline vapors are being enforced in two phases. Phase I requires recovery of the vapor when making a delivery to an underground tank. Phase II, which is currently being implemented in some sections of the United States, requires the recovery of gasoline vapors when an automobile is being fueled. Phase II involves the capture of vapors being expelled from the automobile fuel tank as the liquid fills the tank and displaces the vapors.

At present, there are two types of systems being used for vapor capture at automotive service stations for Phase II: (1) the vapor recovery system, and (2) the vapor processing system. For each system, the *Code* specifies equipment requirements and regulates vapor discharge. (See Figure 7-2.)

The vapor recovery system captures, and retains without processing, flammable liquid vapors displaced during the filling of the tanks or containers, or during the fueling of vehicles. This can be accomplished through systems such as balanced pressure displacement systems, or the so-called hybrid systems. Balanced pressure displacement systems return the gasoline vapors through vapor return hoses and piping back to the underground tanks. The hybrid systems return some of the gasoline through vapor return lines at the pumps before the gasoline reaches the meters. This action draws a vacuum on the gasoline tank being filled, causing vapors to return through the vapor return line.

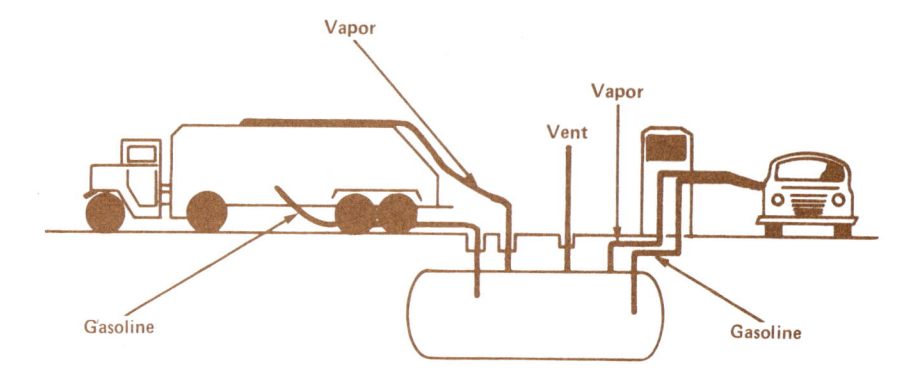

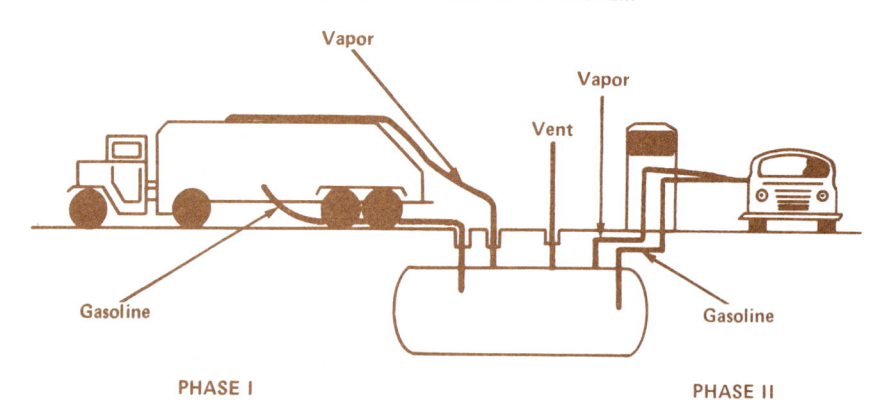

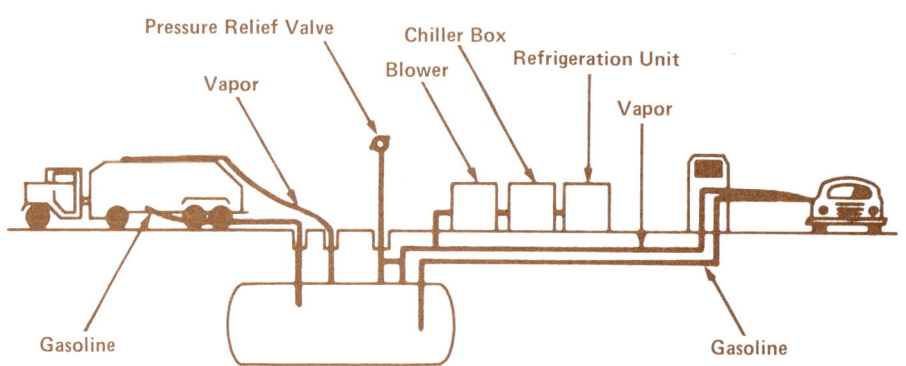

182 FLAMMABLE AND COMBUSTIBLE LIQUIDS CODE HANDBOOK

Figure 7-2. (Top) Vapor Control. (Center) Vapor Recovery System. (Bottom) Vapor Processing System.

7-4.4.1 Dispensing devices incorporating provisions for vapor recovery shall be listed.

7-4.4.2 Existing listed or labeled dispensing devices may be modified for vapor recovery provided that the modifications made are "Listed by Report" by an approved testing laboratory. The "Listing by Report" shall contain a description of the component parts used in the modifications and the recommended method of installation on specific dispensers and it shall be made available on request to the authority having jurisdiction.

"Listed by Report" is a special form of listing employed by Underwriters Laboratories Inc. which covers products or construction for which there are no generally recognized installation requirements. Information concerning proper field installation is contained in a report identified by the reference and date shown by the listing, copies of which may be obtained from the listed company.

7-4.4.3 Hose nozzle valves used on vapor recovery systems shall be listed.

7-4.4.4 Means shall be provided in the vapor return path from each dispensing outlet to prevent the discharge of vapors when the hoze nozzle valve is in its normal non-dispensing position.

7-4.5 Vapor Processing Systems.

Vapor processing captures and processes the flammable liquid vapors displaced during the filling of tanks or containers or when fueling vehicles. The vapor is captured by use of mechanical and/or chemical means through systems that use blower-assist for capturing vapors, and refrigeration, adsorption, or combustion for processing the vapors. Generally, those systems using refrigeration, and in some cases adsorption, are found in large bulk plants. The systems require the use of listed dispensing devices. Restrictions relative to modifying the dispensing devices and the prevention of vapor discharge when the dispenser is not in use are the same for both the vapor processing system and the vapor recovery system. Operators are not permitted to modify listed vapor recovery and/or processing systems.

7-4.5.1 Vapor processing system components consisting of hose nozzle valves, blowers or vacuum pumps, flame arresters or systems for prevention of flame propagation, controls, and vapor processing equipment shall be individually listed for use in a specified manner.

7-4.5.2 Dispensing devices used with a vapor processing system shall be listed. Existing listed or labeled dispensing devices may be modified for use with vapor processing systems provided they are "Listed by Report" as specified in 7-4.4.2.

7-4.5.3 Means shall be provided in the vapor return path from each dispensing outlet to prevent the discharge of vapors when the hose nozzle valve is in its normal non-dispensing position.

7-4.5.4 Vapor processing systems employing blower-assist shall not be used unless the system is designed to prevent flame propagation through system piping, processing equipment and tanks.

Systems employing blower-assist are to be designed to prevent flame propagations through the system piping, processing equipment, and tanks. This additional design requirement is necessary because the introduction of air by use of the blower can dilute the vapors in the piping and bring them into the flammable range.

7-4.5.5 If a component is likely to contain a flammable vapor-air mixture under operating conditions, and can fail in a manner to ignite the mixture, it shall be designed to withstand an internal explosion without failure to the outside.

For example, a fan used in a blower-assist operation may overheat because of a bad bearing, or may cause sparks by rubbing the fan against the housing. Since either of these could provide a possible source of ignition, the housing for the fan must be designed to withstand a possible explosion if there may also be unusual air leakage into the system. When a vapor processing system component is located in an area where the public can be exposed to the hazards of an exploding fragment, such as on a pump island, this requirement assumes added significance.

7-4.5.6 Vapor processing equipment shall be located outside of buildings at least 10 ft (3.05 m) from adjacent property lines which can be built upon, except as provided for in 7-4.5.7. Vapor processing equipment shall be located a minimum of 20 ft (6.1 m) from dispensing devices. Processing equipment shall be protected against physical damage by the provision of guard rails, curbs, or fencing.

7-4.5.7 Where the required distance to adjacent property lines which can be built upon as specified in 7-4.5.6 cannot be obtained, means shall be provided to protect vapor processing equipment against fire exposure. Such means may include protective enclosures which extend at least 18 in. (457.2 mm) above the equipment, constructed of fire resistant or noncombustible materials, installation in below-grade spaces, or protection with an approved water spray system. If protective enclosures or below-grade spaces are used, positive means shall be provided to ventilate the volume within the enclosure to prevent pocketing of flammable vapors. In no case shall vapor processing equipment so protected be located within 5 ft (1.5 m) of adjacent property lines which can be built upon.

7-4.5.8 Electrical equipment shall be in accordance with Table 7-5.

7-4.5.9 Vents on vapor processing systems shall be not less than 12 ft (3.6 m) above adjacent ground level, with outlets so directed and located that flammable vapors will not accumulate or travel to an unsafe location or enter buildings.

7-4.5.10 Combustion or open flame type devices shall not be installed in a classified area. See Table 7-5.

On the other hand, where an open flame device is an inherent part of a process which cannot be safeguarded from vapor access, it is obviously unnecessary to place restrictions on the type of electrical equipment to be used in this area.

7-5 Service Stations Located Inside Buildings.

This entire section is new to the 1981 edition of the *Code*. For some time, an apparent conflict existed between NFPA 30 and NFPA 88A, *Standard for Parking Garages*. The latter code restricted dispensing of fuel to grade level locations in parking garages. NFPA 30, on the other hand, made mention of dispensing units inside buildings, but left the determination for location of such units up to the authority having jurisdiction. This new section resolves the apparent discrepancy and also offers more detailed guidance on the placement of dispensing units inside buildings. It is anticipated that, in its next edition, NFPA 88A will delete all provisions dealing with dispensing unit location and that the code user will be referred to NFPA 30 for guidance on the matter.

7-5.1 General.

7-5.1.1 A service station is permitted inside a building subject to approval of the authority having jurisdiction.

7-5.1.2 The service station shall be separated from other portions of the building by wall, partition, floor, or floor-ceiling assemblies having a fire resistance rating of not less than 2 hr.

7-5.1.3 Interior finish of service stations shall be constructed of noncombustible or approved limited-combustible materials.

7-5.1.4 Door and window openings in interior walls shall be provided with listed 1½-hr (B) fire doors. Doors shall be self-closing, or may remain open during normal operations if they are designed to close automatically in a fire emergency by provision of listed closure devices. Fire doors shall be installed in accordance with NFPA 80, *Standard for Fire Doors and Windows*.

7-5.1.5 Fire doors shall be kept unobstructed at all times. Appropriate signs and markings shall be used.

7-5.1.6 Openings in interior partitions and walls for ducts shall be protected by listed fire dampers. Openings in floor or floor-ceiling assemblies for ducts shall be protected with enclosed shafts. Enclosure of shafts shall be with wall or partition assemblies having a fire resistance rating of not less than 2 hr. Openings in enclosed shafts, for ducts, shall be protected with listed fire dampers.

186   FLAMMABLE AND COMBUSTIBLE LIQUIDS CODE HANDBOOK

7-5.2 Dispensing Area.

7-5.2.1 The dispensing area shall be located at street level, with no dispenser located more than 50 ft (15.24 m) from the vehicle exit to, or entrance from, the outside of the building.

7-5.2.2 Dispensing shall be limited to the area required to serve not more than four vehicles at one time.

7-5.3 Ventilation.

7-5.3.1 Forced air heating, air conditioning, and ventilating systems serving the service station area shall not be interconnected with any such systems serving other parts of the building. Such systems shall be installed in accordance with the provisions of NFPA 90A, *Standard for the Installation of Air Conditioning and Ventilating Systems*.

7-5.3.2 A mechanical exhaust system shall be provided to serve only the dispensing area. This system shall be interlocked with the dispensing system such that air flow is established before any dispensing unit can operate. Failure of air flow shall automatically shut down the dispensing system.

7-5.3.3 The exhaust system shall be designed to provide air movement across all portions of the dispensing area floor, and to prevent the flow of flammable vapors beyond the dispensing area. Exhaust inlet ducts shall not be less than 3 in. (76.2 mm) nor more than 12 in. (304.8 mm) above the floor. Exhaust ducts shall not be located in floors, or penetrate the floor of the dispensing area, and shall discharge to a safe location outside the building.

7-5.3.4 The exhaust system shall provide ventilation at a rate of not less than 1 cu ft (0.28 m^3) per minute per sq ft (0.0929 m^2) of dispensing area.

7-5.3.5 Exhaust system shall be installed in accordance with the provisions of NFPA 91, *Standard for Blower and Exhaust Systems*.

7-5.3.6 The provisions of 7-5.3.2, 7-5.3.3, 7-5.3.4 and 7-5.3.5 do not apply to a service station located inside a building if 2 or more sides of the dispensing area are open to the building exterior such that natural ventilation can normally be expected to dissipate flammable vapors.

7-5.4 Piping

7-5.4.1 Piping systems shall comply with the provisions of Chapter 3.

7-5.4.2 All fuel and flammable vapor piping inside buildings but outside the service station area shall be enclosed within a horizontal chase or a vertical shaft used only for this piping. Vertical shafts and horizontal chases shall be constructed of materials having a fire resistance rating of not less than 2 hr.

7-5.5 Drainage Systems.

7-5.5.1 Floors shall be liquidtight. Emergency drainage systems shall be provided to direct flammable or combustible liquid leakage and fire protection water to a safe location. This may require curbs, scuppers, or special drainage systems.

7-5.5.2 Emergency drainage systems, if connected to public sewers or discharged into public waterways, shall be equipped with traps or separators.

7-6 Electrical Equipment.

7-6.1 Section 7-6 shall apply to areas where Class I liquids are stored, handled or dispensed. For areas where Class II or Class III liquids are stored, handled or dispensed, the electrical equipment may be installed in accordance with the provisions of NFPA 70, *National Electrical Code*, for nonclassified locations.

7-6.2 All electrical equipment and wiring shall be of a type specified by and shall be installed in accordance with NFPA 70, *National Electrical Code*. All electrical equipment integral with the dispensing hose or nozzle shall be suitable for use in Division 1 locations.

> **In addition to those provisions which are the same for all occupancies, the *Code* specifically requires that all electrical equipment integral to the dispensing hose or nozzle must be suitable for use in Class I, Division 1 locations.**

7-6.3 Table 7-5 shall be used to delineate and classify areas for the purpose of installation of electrical equipment under normal circumstances. A classified area shall not extend beyond an unpierced wall, roof or other solid partition. The designation of classes and divisions is defined in Chapter 5, Article 500, of NFPA 70, *National Electrical Code*.

> **Note that Table 7-5 limits the type of permanently installed electrical wiring and equipment that may be placed in the specified areas. It is not applicable or intended to restrict the presence of other nonelectric equipment such as the hot surfaces on an automobile.**

7-6.4 The area classifications listed in 7-6.3 shall be based on the premise that the installation meets the applicable requirements of this code in all respects. Should this not be the case, the authority having jurisdiction shall have the authority to determine the extent of the classified area.

7-7 Heating Equipment.

> **Heating equipment can be a source of ignition of flammable vapors in service stations if installed so that ignitible vapor-air mixtures can reach it.**

7-7.1 Heating equipment shall be installed as provided in 7-7.2 through 7-7.6.

Table 7-5

Electrical Equipment Classified Areas — Service Stations

Location	NEC Class I, Group D Division	Extent of Classified Area
UNDERGROUND TANK		
Fill Opening	1	Any pit, box or space below grade level, any part of which is within the Division 1 or 2 classified area.
	2	Up to 18 inches above grade level within a horizontal radius of 10 feet from a loose fill connection and within a horizontal radius of 5 feet from a tight fill connection.
Vent — Discharging Upward	1	Within 3 feet of open end of vent, extending in all directions.
	2	Area between 3 feet and 5 feet of open end of vent, extending in all directions.
DISPENSING UNITS (except overhead type)		
Pits	1	Any pit, box or space below grade level, any part of which is within the Division 1 or 2 classified area.
Dispenser	1	The area within a dispenser enclosure up to 4 feet vertically above the base except that area defined as Division 2. Any area within a nozzle boot.
	2	Areas within a dispenser enclosure above the Division 1 area. Areas within a dispenser enclosure isolated from Division 1 by a solid partition or a solid nozzle boot but not completely surrounded by Division 1 area. Within 18 inches horizontally in all directions from the Division 1 area located within the dispenser enclosure. Within 18 inches horizontally in all directions from the opening of a nozzle boot not isolated by a vapor-tight partition, except that the classified area need not be extended around a 90° or greater corner.

Table 7-5 — *continued*

Location	NEC Class I, Group D Division	Extent of Classified Area
Outdoor	2	Up to 18 inches above grade level within 20 feet horizontally of any edge of enclosure.
INDOOR with Mechanical Ventilation	2	Up to 18 inches above grade or floor level within 20 feet horizontally of any edge of enclosure.
with Gravity Ventilation	2	Up to 18 inches above grade or floor level within 25 feet horizontally of any edge of enclosure.
DISPENSING UNITS, OVERHEAD TYPE	1	Within the dispenser enclosure and 18 inches in all directions from the enclosure where not suitably cut off by ceiling or wall. All electrical equipment integral with the dispensing hose or nozzle.
	2	An area extending 2 feet horizontally in all directions beyond the Division 1 area and extending to grade below this classified area.
	2	Up to 18 inches above grade level within 20 feet horizontally measured from a point vertically below the edge of any dispenser enclosure.
REMOTE PUMP — OUTDOOR	1	Any pit, box or space below grade level if any part is within a horizontal distance of 10 feet from any edge of pump.
	2	Within 3 feet of any edge of pump, extending in all directions. Also up to 18 inches above grade level within 10 feet horizontally from any edge of pump.
REMOTE PUMP — INDOOR	1	Entire area within any pit.
	2	Within 5 feet of any edge of pump, extending in all directions. Also up to 3 feet above floor or grade level within 25 feet horizontally from any edge of pump.

Table 7-5 — *continued*

Location	NEC Class I, Group D Division	Extent of Classified Area
LUBRICATION OR SERVICE ROOM — with Dispensing	1	Any pit within any unventilated area.
	2	Any pit with ventilation.
	2	Area up to 18 inches above floor or grade level and 3 feet horizontally from a lubrication pit.
Dispenser for Class I Liquids	2	Within 3 feet of any fill or dispensing point, extending in all directions.
LUBRICATION OR SERVICE ROOM — WITHOUT DISPENSING	2	Entire area within any pit used for lubrication or similar services where Class I liquids may be released.
	2	Area up to 18 inches above any such pit, and extending a distance of 3 feet horizontally from any edge of the pit.
SPECIAL ENCLOSURE INSIDE BUILDING PER 7-2.2	1	Entire enclosure.
SALES, STORAGE AND REST ROOMS	Non-classified	If there is any opening to these rooms within the extent of a Division 1 area, the entire room shall be classified as Division 1.
VAPOR PROCESSING SYSTEMS PITS	1	Any pit, box or space below grade level, any part of which is within a Division 1 or 2 classified area or which houses any equipment used to transport or process vapors.
VAPOR PROCESSING EQUIPMENT LOCATED WITHIN PROTECTIVE ENCLOSURES (See 7-4.5.7)	2	Within any protective enclosure housing vapor processing equipment.

Table 7-5 — *continued*

Location	NEC Class I, Group D Division	Extent of Classified Area
VAPOR PROCESSING EQUIPMENT NOT WITHIN PROTECTIVE ENCLOSURES (excluding piping and combustion devices)	2	The space within 18 inches in all directions of equipment containing flammable vapor or liquid extending to grade level. Up to 18 inches above grade level within 10 ft. horizontally of the vapor processing equipment.
EQUIPMENT ENCLOSURES	1	Any area within the enclosure where vapor or liquid is present under normal operating conditions.
	2	The entire area within the enclosure other than Division 1.
VACUUM ASSIST BLOWERS	2	The space within 18 inches in all directions extending to grade level. Up to 18 inches above grade level within 10 feet horizontally.

SI Units: 1 in. = 25.40 mm; 1 ft = 0.3048 m.

7-7.2 Heating equipment may be installed in the conventional manner except as provided in 7-7.3, 7-7.4, 7-7.5, or 7-7.6.

Heating equipment installation is to be accomplished in the same manner as for an ordinary or general purpose area, except as modified by 7-7.3 through 7-7.6.

7-7.3 Heating equipment may be installed in a special room separated from an area classified as Division 1 or Division 2 in Table 7-5 by walls having a fire resistance rating of at least 1 hr and without any openings in the walls within 8 ft (2.4 m) of the floor into an area classified as Division 1 or Division 2 in Table 7-5. This room shall not be used for combustible storage, and all air for combustion purposes shall come from outside the building.

The opening at the top of the wall allows heated air to enter the service area, but the 8 ft (2.4 m) wall height is high enough to keep flammable vapors from passing into a special room. The air for combustion purposes is that required to supply the combustion chamber of the heating unit. This air inlet shall be carefully located to avoid the flow of flammable liquid vapors into the heating unit.

7-7.4 Heating equipment using gas or oil fuel may be installed in the lubrication or service room where there is no dispensing or transferring of Class I liquids provided the bottom of the combustion chamber is at least 18 in. (457.2 mm) above the floor and the heating equipment is protected from physical damage.

> Generally, flammable vapors should not be present in a service station building. However, the 18 in. (457.2 mm) are required as protection against inadvertently providing a source of ignition for any flammable vapors that may be at the floor level due to leakage or spillage during automobile repair or any other activities within the building. Under no circumstances should service station or garage attendants use Class I flammable liquids as cleaning agents for either tools or floor.
>
> Note that gas or oil fuel is specified. The question arises as to why coal or wood burning devices are not allowed. The answer lies in the problem of live ashes or embers. Even if the ash pit is 18 in. (457.2 mm) above the floor, live particles might constitute a source of ignition when they are being removed from the heating unit.

7-7.5 Heating equipment using gas or oil fuel listed for use in garages may be installed in the lubrication or service room where Class I liquids are dispensed provided the equipment is installed at least 8 ft (2.4 m) above the floor.

> Spilled Class I liquid will release vapor at floor level. The vapor will not rapidly diffuse upward, and the 8 ft (2.4 m) height limit guards against the chance that the heater will become an ignition source for the vapor.

7-7.6 Electrical heating equipment shall conform to Section 7-6.

7-8 Operational Requirements.

7-8.1 Fuel Delivery Nozzles.

7-8.1.1 A listed automatic-closing type hose nozzle valve, with or without latch-open device, shall be provided on island-type dispensers used for the dispensing of Class I liquids.

> The words "with or without latch-open device" are new to the 1981 edition of the *Code*. Prior to this change, latch-open devices were not permitted on nozzles unless the dispensing was performed by the service station attendant. At self-service dispensing, where the public had access to the nozzle, the latch-open devices were prohibited.
>
> There are several reasons behind the change in the 1981 edition. Enforcement agencies advised that many citizens chocked open the nozzles with whatever device was at hand, including the gas cap from the automobile. Some enterprising individuals even marketed a combination key holder and latch-open device for one's personal use. Enforcement of the ban on latch-open devices proved extremely difficult. In locations where latch-open devices were allowed by

authorities having jurisdiction, no untoward incidents were reported to NFPA. Finally, several consumers complained that the prohibition of the latch-open device inconvenienced the consumer, questioned his intelligence, and gave more credit for common sense and for concern for safety to the service station attendant than to the public at large.

For all these reasons, the change was made. By using the wording "with or without," the final decision falls to the local authority having jurisdiction. The Technical Committee decided that there is insufficient justification for continuing the outright ban on the use of the latch-open device.

7-8.1.2 If a hose nozzle valve is provided with a latch-open device other than recommended by the valve manufacturer, the valve shall conform to the applicable requirements of Section 19 of UL 842-1974, *Standard for Valves for Flammable Fluids*.

7-8.1.3 Overhead-type dispensing units shall be provided with a listed automatic-closing type hose nozzle valve without a latch-open device.

(a) A listed automatic-closing type hose nozzle valve with latch-open device may be used if the design of the system is such that the hose nozzle valve will close automatically in the event the valve is released from a fill opening or upon impact with a driveway.

7-8.1.4 Dispensing nozzles used at marine service stations shall be of the automatic-closing type without a latch-open device.

The restriction on latch-open devices remains in effect at marine service stations. The rationale used in maintaining this requirement is that the overfilling of a boat's fuel tank results in a quite serious condition. Gasoline spilled into the bilge or interior of a boat can be much more hazardous than a similar spill on the surface of a service station. Aboard a marine vessel, the vapors will be more confined, and dispersal of the vapors will take a longer period of time. Control of sources of ignition may also be more difficult. The absence of a latch-open device requires greater attention, on the part of the individual performing the dispensing operation, to the task at hand.

7-8.1.5 A hose nozzle valve used for dispensing Class I liquids into a container shall be manually held open during the dispensing operation.

7-8.2 Dispensing into Portable Containers. No delivery of any Class I or Class II liquid shall be made into portable containers unless the container is constructed of metal or is approved by the authority having jurisdiction, has a tight closure and is fitted with spout or so designed that the contents can be poured without spilling.

Portable containers of flammable and combustible liquids may be subjected to a great deal of physical abuse because of their portable nature. They also may be used or carried near a wide variety of sources of ignition. Therefore, it is important that the container be constructed to withstand impacts. The possibility of leakage and escaping vapors must also be controlled. Containers should also be designed to minimize spillage when the contents are being transferred.

7-8.2.1 No sale or purchase of any Class I, Class II or Class III liquids shall be made in containers unless such containers are clearly marked with the name of the product contained therein.

Because all of the contents of a portable container may not be used at the time they are purchased, it is important that the container be clearly marked with the name of the product contents in an effort to avoid potential problems and difficulties at a later date. This also assists the station operator by permitting him to refuse to dispense into an unsuitable container.

7-8.3 Attendance or Supervision of Dispensing.

The provisions of 7-8.3.4 and 8-7.3.5 are more easily understood if they are treated as exceptions to the basic provisions on the storage of liquids in service stations.

7-8.3.1 Each service station open to the public shall have an attendant or supervisor on duty whenever the station is open for business.

7-8.3.2 Listed "self-service" dispensing devices are permitted at service stations available and open to the public provided that all dispensing of Class I liquids by a person other than the service station attendant is under the supervision and control of a qualified attendant.

7-8.3.3 Dispensing of liquids at private locations, where the dispensing equipment is not open to the public, does not require an attendant or supervisor. Such locations may include card or key controlled dispensers.

Originally a Tentative Interim Amendment (TIA) to the 1977 edition of the *Code*, this paragraph has now become part of the actual text. No comments were received on the proposed insertion of the TIA to the 1981 edition.

Private stations, or stations not open to the general public, have been in existence for many years. The *Code* first addressed them in 1966. There has been no adverse fire experience at these installations that for years have utilized card or key operated dispensing devices. The expanded use of this type of station, beyond the commercial, industrial, governmental or manufacturing establishments, has been as safe as other types of service stations. The Technical Committee felt that restrictions on card or key operated devices which are located at stations not open to the public could not be justified. Ample time had been provided for public scrutiny and input on the subject, but no comments were received.

7-8.3.4 The provisions of 7-2.1.1 shall not prohibit the temporary use of movable tanks in conjunction with the dispensing of flammable or combustible liquids into the fuel tanks of motor vehicles or other motorized equipment on premises not normally accessible to the public. Such installations shall only be made under permit from the enforcing authority. The permit shall include a definite time limit.

7-8.3.5 The provisions of 7-2.1.1 shall not prohibit the dispensing of Class I and Class II liquids in the open from a tank vehicle to a motor vehicle. Such dispensing shall be permitted provided:

(a) An inspection of the premises and operations has been made and approval granted by the authority having jurisdiction.

(b) The tank vehicle complies with the requirements covered in NFPA 385, *Recommended Regulatory Standard for Tank Vehicles for Flammable and Combustible Liquids*.

(c) The dispensing is done on premises not open to the public.

(d) The dispensing hose does not exceed 50 ft (15.24 m) in length.

(e) The dispensing nozzle is a listed automatic-closing type without a latch-open device.

(f) Nighttime deliveries shall only be made in adequately lighted areas.

(g) The tank vehicle flasher lights shall be in operation while dispensing.

(h) Fuel expansion space shall be left in each fuel tank to prevent overflow in the event of temperature increase.

> These provisions allow groups working on construction sites to have their own fuel supply for motorized construction equipment. Another application of this exception is the possible fueling of motor vehicles from a tank vehicle at large parking lots with access limited to employees of the user of the lot. Note that this provision does not authorize the use of elevated skid tanks which are commonly seen at construction sites. Paragraph 7-8.3.5 applies only to dispensing from a *tank vehicle*.

7-8.4 Self-Service Stations.

7-8.4.1 Self-service station shall mean that portion of property where liquids used as motor fuels are stored and subsequently dispensed from fixed approved dispensing equipment into the fuel tanks of motor vehicles by persons other than the service station attendant, and may include facilities available for sale of other retail products.

> Originally, the public fire services objected strongly to self-service stations,

and such operations were prohibited by NFPA 30. The Fire Marshals Association of North America reviewed the objections and appointed a subcommittee composed of state, county, and city fire marshals to develop a code requirement to regulate self-service stations. The subcommittee recognized that private service stations had a reasonably good fire record and that self-service stations had been used for many years in several locations in this country, apparently without bad experiences. The Technical Committee responsible for NFPA 30 adopted the requirements recommended by the Fire Marshals' subcommittee, and subsequently expanded them to include the convenience stores that also dispense gasoline in front of the store building.

Based on data received and developed over the last several years, it is impossible to make a determination regarding the relative safety of the self-service versus the full-service station. It cannot be stated that one is more or less hazardous than the other. Quite often, the nature of the fire incident has no relationship to the type of service station. For example, an occurrence which involves a vehicle knocking over a dispensing unit can happen at either a full-service or self-service station.

7-8.4.2 Listed dispensing devices such as, but not limited to, coin-operated, card-operated and remote controlled types are permitted at self-service stations.

7-8.4.3 All self-service stations shall have at least one attendant on duty while the station is open to the public. The attendant's primary function shall be to supervise, observe and control the dispensing of Class I liquids while said liquids are actually being dispensed.

> The attendant's primary function is to supervise the dispensing of Class I liquids during the actual filling operation. This would include preventing the dispensing of Class I liquids into portable containers which are not approved for such storage.

7-8.4.4 It shall be the responsibility of the attendant to (1) prevent the dispensing of Class I liquids into portable containers not in compliance with 7-8.2; (2) control sources of ignition; and (3) immediately handle accidental spills and fire extinguishers if needed. The attendant or supervisor on duty shall be mentally and physically capable of performing the functions and assuming the responsibility prescribed in this section.

> As with regulations governing other occupancies, the chapter on service stations deals with the problems of sources of ignition and fire control. By eliminating or controlling the major sources of ignition such as smoking or open flames in the areas used for fueling, receiving or dispensing Class I liquids, and shutting off all motors being fueled, hazards are held to an acceptable minimum.

7-8.4.5 Emergency controls specified in 7-4.2 shall be installed at a location acceptable to the authority having jurisdiction, but controls shall not be more than 100 ft (30.48 m) from dispensers.

7-8.4.6 Operating instructions shall be conspicuously posted in the dispensing area.

7-8.4.7 The dispensing area shall at all times be in clear view of the attendant, and the placing or allowing of any obstacle to come between the dispensing area and the attendant control area shall be prohibited. The attendant shall at all times be able to communicate with persons in the dispensing area.

7-8.4.8 Warning signs shall be conspicuously posted in the dispensing area incorporating the following or equivalent wording: (a) WARNING — It is unlawful and dangerous to dispense gasoline into unapproved containers; (b) No Smoking; and (c) Stop Motor.

7-8.5 Drainage and Waste Disposal.

7-8.5.1 Provision shall be made in the area where Class I liquids are dispensed to prevent spilled liquids from flowing into the interior of service station buildings. Such provision may be made by grading driveways, raising door sills, or other equally effective means.

7-8.5.2 Crankcase drainings and liquids shall not be dumped into sewers, streams or adjoining property, but shall be stored in tanks or drums outside any building until removed from the premises.

Small tanks or drums are often used for the storage of crankcase drainings and similar liquids. One must be aware of the possibility that these waste oils may be contaminated with lower flash point flammable liquids. Thus, waste oils are not to be overlooked in applying regulations. For environmental and safety reasons, crankcase drainings should not be dumped into sewers.

7-9 Sources of Ignition.

7-9.1 In addition to the previous restrictions of this chapter, the following shall apply: There shall be no smoking or open flames in the areas used for fueling, servicing fuel systems for internal combustion engines, or receiving or dispensing of Class I liquids. Conspicuous and legible signs prohibiting smoking shall be posted within sight of the customer being served. The motors of all equipment being fueled shall be shut off during the fueling operation except for emergency generators, pumps, etc., where continuing operation is essential.

7-10 Fire Control.

7-10.1 Each service station shall be provided with at least one listed fire extinguisher having a minimum classification of 5B:C located so that an extinguisher will be within 100 ft (30.48 m) of each pump, dispenser, underground fill pipe opening, and lubrication or service room.

REFERENCES CITED BY *CODE*

(These publications comprise a part of the requirements to the extent called for by the **Code.***)*

NFPA 385, *Tank Vehicles for Flammable and Combustible Liquids,* **NFPA, Boston, 1979.**

NFPA 70, *National Electrical Code,* **NFPA, Boston, 1981.**

NFPA 80, *Fire Doors and Windows,* **NFPA, Boston, 1979.**

NFPA 90A, *Air Conditioning and Ventilating Systems,* **NFPA, Boston, 1978.**

NFPA 91, *Blower and Exhaust Systems, Dust, Stock & Vapor Removal or Conveying,* **NFPA, Boston, 1973.**

NFPA 302, *Fire Protection Standard for Motor Craft,* **NFPA, Boston, 1980.**

NFPA 303, *Fire Protection Standard for Marinas and Boatyards,* **NFPA, Boston, 1975.**

Underwriters Laboratories Inc., *Standard for Valves for Fluids,* **UL 842, Chicago, 1977.**

American Petroleum Institute, *Recommended Practice for Bulk Liquid Stock Control at Retail Outlets,* **API Publication 1621, 3rd ed., Washington, DC, 1977.**

REFERENCES CITED IN COMMENTARY

[1]**NFPA 329,** *Underground Leakage of Flammable and Combustible Liquids,* **NFPA, Boston, 1977.**

[2]**American Petroleum Institute,** *Recommended Practice for Bulk Liquid Stock Control at Retail Outlets,* **API Publication 1621, 3rd ed., Washington, DC, 1977.**

8

Processing Plants

8-1 Scope.

8-1.1 This chapter shall apply to those plants or buildings which contain chemical operations such as oxidation, reduction, halogenation, hydrogenation, alkylation, polymerization, and other chemical processes but shall not apply to chemical plants, refineries or distilleries as defined and covered in Chapter 9, Refineries, Chemical Plants and Distilleries.

> As mentioned under 5-1.1, this chapter considers the situation where the hazards involved in use of liquids are complicated by the simultaneous presence of chemical reactions with potential for heat release.

8-2 Location.

8-2.1 The location of each processing vessel shall be based upon its liquid capacity. Processing vessels shall be located, with respect to distances to lines of adjoining property which can be built upon, in accordance with Table 8-2.1, except when the processing plant is designed in accordance with 8-2.1.1.

Table 8-2.1

Location of Processing Vessels from Property Lines

Processing Vessels with Emergency Relief Venting to Permit Pressure	Stable Liquids	Unstable Liquids
Not in excess of 2.5 psig (17.2 kPa)	Table 2-6*	2½ times Table 2-6*
Over 2.5 psig (17.2 kPa)	1½ times Table 2-6*	4 times Table 2-6*

*Double distances where protection of exposure is not provided.

8-2.1.1 The distances required in 8-2.1 may be waived when the vessels are housed within a building and the exterior wall facing the line of adjoining property which can be built upon is a blank wall having a fire resistance rating of not less than four hours. When Class IA or unstable liquids are handled, the blank wall shall have explosion resistance in accordance with good engineering practice (*see 8-3.4*).

> As mentioned under 5-3.2, spacing to property that can be built upon is measured from the vessel containing the liquid. If the vessel is in the open, dikes or drainage should be provided as specified in Chapter 2, Tank Storage; if in a building, the walls and the required drainage may provide the necessary isolation, although the use of scuppers (see 8-3.2.1) and explosion relief (see 8-3.4.1) may make a wall unsuitable. The greater spacing required in the case of vessels for unstable liquids is based on the possibility that heat-producing chemical reactions may exceed the capacity of the venting system. Unlike the case in Chapter 5 (see 5-3.3), processing vessels are not required to be separated from the remainder of the plant. The note to Table 8-2.1 refers to fire protection — such as that afforded by a fire department — for property of others that would be exposed by a fire involving the liquid.

8-3 Processing Buildings.

8-3.1 Construction.

8-3.1.1 Processing buildings shall be of fire-resistive or noncombustible construction, except heavy timber construction with load-bearing walls may be permitted for plants utilizing only stable Class II or Class III liquids. Except as provided in 8-2.1.1 or in the case of explosion resistant walls used in conjunction with explosion relieving facilities (*see 8-3.4*), load-bearing walls shall be prohibited. Buildings handling Class I or Class II liquids shall be without basements or covered pits. Processing buildings are normally limited in height and area, depending upon the type of construction and private fire protection provided, to minimize the possibility of fire of such extent as to jeopardize public safety. Processing buildings with numerous accessible exterior openings offer favorable features for fire fighting. Provision for smoke and heat venting may be desirable to assist access for fire fighting. NFPA 204, *Guide for Smoke and Heat Venting*, provides information on this subject.

> Since the consensus is that the explosion and fire hazards are greater in processing plants than in industrial plants, types of construction least susceptible to these hazards are specified. There have been several cases where sudden ignition of vapors from a spill, caused by an uncontrolled chemical reaction, have opened all the sprinkler heads in the area involved. Limitations of height and area should therefore be correlated with the fire control measures. (See Section 8-6.)

8-3.1.2 Areas shall have adequate exit facilities arranged to prevent occupants from being trapped in the event of fire. Exits shall not be exposed by the drainage facilities described in 8-3.2. NFPA 101, *Life Safety Code*, provides information on this subject.

With respect to NFPA 101,[1] the requirements for high hazard occupancies, Section 5-11, should be followed.

8-3.2 Drainage.

8-3.2.1 Emergency drainage systems shall be provided to direct flammable or combustible liquid leakage and fire protection water to a safe location. This may require curbs, scuppers, or special drainage systems to control the spread of fire (*see 2-2.3.1*). Appendix A of NFPA 15, *Standard for Water Spray Fixed Systems for Fire Protection*, provides information on this subject.

8-3.2.2 Emergency drainage systems, if connected to public sewers or discharged into public waterways, shall be equipped with traps or separators.

8-3.2.3 The processing plant shall be designed and operated to prevent the normal discharge of flammable or combustible liquids to public waterways, public sewers, or adjoining property.

These items were discussed under 5-3.4.1, 5-3.4.2, and 5-3.4.3.

8-3.3 Ventilation.

8-3.3.1 Enclosed processing buildings handling Class I or Class II liquids shall be ventilated at a rate of not less than 1 cu ft (0.02832 m^3) per minute per sq ft (0.0929 m^2) of solid floor area. This shall be accomplished by natural or mechanical ventilation with discharge or exhaust to a safe location outside of the building. Provision shall be made for introduction of make-up air in such a manner as not to short circuit the ventilation. Ventilation shall be arranged to include all floor areas or pits where flammable vapors may collect. Local or spot general ventilation may be needed for the control of special fire or health hazards. Such ventilation, if provided, can be utilized for up to 75 percent of the required ventilation. NFPA 91, *Standard for the Installation of Blower and Exhaust Systems for Dust, Stock and Vapor Removal or Conveying*, and NFPA 90A, *Standard for the Installation of Air Conditioning and Ventilating Systems*, provide information on this subject.

This was discussed under 5-3.5.1. However, Class II liquids are included here. This is because many chemical reactions involve elevated temperatures, and vapors from heated Class II liquids can travel a considerable distance before cooling. (See 1-1.3.) Vapors from heated Class III liquids cool much more rapidly because of the higher temperature differential between such vapors and the air.

8-3.3.2 Equipment used in a building and the ventilation of the building shall be designed so as to limit flammable vapor-air mixtures, under normal operating conditions, to the interior of equipment, and to not more than 5 ft (1.5 m) from equipment which exposes Class I liquids to the air. Examples of such equipment are dispensing stations, open centrifuges, plate and frame filters, open vacuum filters, and surfaces of open equipment.

This was discussed under 5-3.5.2.

8-3.4 Explosion Relief.

8-3.4.1 Areas where Class IA or unstable liquids are processed shall have explosion venting through one or more of the following methods: (a) open air construction; (b) lightweight walls and roof; (c) lightweight wall panels and roof hatches; (d) windows of explosion venting type. NFPA 68, *Guide for Explosion Venting*, provides information on this subject.

This is a minimum requirement. Explosion venting is also desirable where heated vapors from other classes of liquids may be released. As hot vapors in the flammable range cool and condense, they can form an explosive cloud or mist.

8-4 Liquid Handling.

Note that by reference in 5-3.6, this section may also apply to industrial plants.

8-4.1 Storage.

8-4.1.1 The storage of liquids in tanks shall be in accordance with the applicable provisions of Chapter 2, Tank Storage.

8-4.1.2 If the storage of liquids in outside aboveground or underground tanks is not practical because of government regulations, temperature considerations or production considerations, tanks may be permitted inside of buildings or structures in accordance with the applicable provisions of Chapter 2, Tank Storage. Production considerations necessitating storage inside of buildings include but are not limited to high viscosity, purity, sterility, hydroscopicity, sensitivity to temperature change, and need to store temporarily pending completion of sample analysis.

Because they constitute a high fire load, liquids should preferably be stored outdoors. This paragraph recognizes that there may be valid reasons for having storage tanks located indoors.

8-4.1.3 Storage tanks inside of buildings shall be permitted only in areas at or above grade which have adequate drainage and are separated from the processing area by construction having a fire resistance rating of at least 2 hrs. Day tanks, running tanks and surge tanks are permitted in process areas. Openings to other rooms or buildings shall be provided with noncombustible liquidtight raised sills or ramps at least 4 in. (101.6 mm) in height, or the floor in the storage area shall be at least 4 in. (101.6 mm) below the surrounding floor. As a minimum, each opening shall be provided with a listed, self-closing 1½-hr (B) fire door installed in accordance with NFPA 80, *Standard for Fire Doors and Windows*, or a listed fire damper installed where required by NFPA 90A, *Standard for Air Conditioning and Ventilating Systems*, or NFPA 91, *Standard for Blower and Exhaust Systems*. The room shall be liquidtight where the walls join the floor.

> These requirements are self-explanatory, except that drainage is required to be adequate. The intent is that any accidental spill will drain away before overflowing the required barrier. One approach is to make the barrier high enough or the floor depression deep enough so spillage of the contents of the largest tank would not overflow into adjacent areas. Another is to estimate the size of the largest probable leak, and then design the drainage system to handle such a leak without permitting any other flow from the room. Such a design criterion should be reached jointly with the authority having jurisdiction. Note that in theory, in a room with all openings sealed off by fire doors or dampers, a fire will become much less severe and may self-extinguish because of lack of oxygen.[2] However, in practice, by the time the fire has burned long enough to self-extinguish from lack of oxygen, there is bound to be a hot residue of ash and coals which will rekindle the fire when air is admitted.
>
> The permission for use of day tanks, running tanks, and surge tanks is to allow enough liquid for a single batch, or for use during one shift or one day, to be present in the process areas without being separated by a fire wall. In contrast with storage areas, process areas will be continuously occupied. Many batch processes are dependent on the use of large, precisely measured quantities of liquids added at rates indicated by the rate or state of the reaction involved. In such cases, use of remote pumping may result in loss of control of the reaction with a greater probability of serious adverse consequences than the probability of accident as a result of having the liquid in the area in the permitted tanks.

8-4.1.4 The storage of liquids in containers shall be in accordance with the applicable provisions of Chapter 4, Container and Portable Tank Storage.

> Obviously, 4-5.2, 4-5.3, 4-5.4, and 4-5.5 do not apply. Also, note that while Table 4-6.1(a) permits drum storage of Class IB and IC liquids stacked two high, Table C-4-6.2(a) (in Appendix C) gives no advice on sprinkler protection for such storage. This is because of the lack of adequate test data on the matter.

8-4.2 Piping, Valves and Fittings.

8-4.2.1 Piping, valves and fittings shall be in accordance with Chapter 3, Piping, Valves and Fittings.

8-4.2.2 Listed flexible connectors may be used where vibration exists or where frequent movement is necessary. Approved hose may be used at transfer stations.

Note that flexible connectors must be listed, but that hose need only be approved.

8-4.2.3 Piping containing liquids shall be identified.

There are various conflicting methods of color coding for piping. Identification by use of labels or tags may also be used. Whatever the method, consistency throughout the premises is important.

8-4.3 Transfer.

8-4.3.1 The transfer of large quantities of liquids shall be through piping by means of pumps or water displacement. Except as required in process equipment, gravity flow shall not be used. The use of compressed air as a transferring medium shall be prohibited.

This requirement is intended to prohibit the transfer of liquids into vessels by pouring from drums or other containers. The indefinite expression "large quantities" recognizes that some liquids that are used infrequently or are of unusual composition (including exceptional purity) may not be available in bulk.

8-4.3.2 Positive displacement pumps shall be provided with pressure relief discharging back to the tank or to pump suction.

This requirement is designed to prevent a failure because of overpressure in that part of a pumping system which may be plugged or valved off and into which the pump may discharge. In the case of centrifugal pumps (not mentioned in this paragraph), pumping into a closed system or continuous circulation through a bypass may cause heating, a vapor lock, and destructive failure of the pump with consequent release of liquid.

8-4.4 Equipment.

8-4.4.1 Equipment shall be designed and arranged to prevent the unintentional escape of liquids and vapors and to minimize the quantity escaping in the event of accidental release.

8-4.4.2 Where the vapor space of equipment is usually within the flammable range, the probability of explosion damage to the equipment can be limited by inerting, by providing an explosion suppression system, or by designing the equipment to contain the peak explosion pressure which can be modified by explosion relief. Where the special hazards of operation, sources of ignition, or exposures indicate a need, consideration shall be given to providing protection by one or more of the above means. NFPA 69, *Explosion Prevention Systems*, provides information on inerting.

Also, NFPA 68[3] gives advice on the explosion venting phase of protection. Although not mentioned in this paragraph of the *Code*, emergency venting to care for runaway reactions may be needed. An empirical guide to sizing such venting systems is given in the 14th Edition of the NFPA *Fire Protection Handbook*,[4] p. 4-51.

8-5 Tank Vehicle and Tank Car Loading and Unloading.

8-5.1 Tank vehicle and tank car loading or unloading facilities shall be separated from aboveground tanks, warehouses, other plant buildings or nearest line of adjoining property which can be built upon by a distance of 25 ft (7.62 m) for Class I liquids and 15 ft (4.57 m) for Class II and Class III liquids measured from the nearest position of any fill stem. Buildings for pumps or shelters for personnel may be a part of the facility. Operations of the facility shall comply with the appropriate portions of Section 6-3 of Chapter 6, Bulk Plants.

This item is the same as 5-4.1 and is discussed under Section 6-3.

8-6 Fire Control.

This section is similar in intent to Section 5-5. However, the wording is different.

8-6.1 Listed portable fire extinguishers of appropriate size, type and number shall be provided. NFPA 10, *Standard for Portable Fire Extinguishers*, provides information on this subject.

This requires portable fire extinguishers, where 5-5.1 could be satisfied by using small hose connections and water buckets.

8-6.2 Where the special hazards of operation or exposure indicate a need, the following fire control provisions shall be provided.

This paragraph implies consultation with the authority having jurisdiction.

8-6.2.1 A reliable water supply shall be available in pressure and quantity adequate to meet the probable fire demands.

8-6.2.2 Hydrants shall be provided in accordance with accepted good practice. NFPA 24, *Standard for Outside Protection*, provides information on this subject.

8-6.2.3 Hose connected to a source of water shall be installed so that all vessels, pumps, and other equipment containing flammable or combustible liquids can be reached with at least one hose stream. Nozzles that are capable of discharging a water spray shall be provided. NFPA 14, *Standard for the Installation of Standpipes and Hose Systems*, provides information on this subject.

The requirement that a water spray nozzle be provided comes about because: (1) sprays can be used on hot liquids without forcing large droplets below the hot surface and causing spattering, (2) sprays can provide a cooling shield so the fire may be attacked from a more advantageous position, and (3) sprays entrain air and act to move hot gases or unburned vapors in a direction advantageous to the nozzle operator.

8-6.2.4 Processing plants shall be protected by an approved automatic sprinkler system or equivalent extinguishing system. If special extinguishing systems including but not limited to those employing foam, carbon dioxide or dry chemical are provided, listed equipment shall be used and installed in accordance with NFPA 11, *Standard for Foam Extinguishing Systems*; NFPA 12, *Standard on Carbon Dioxide Extinguishing Systems*; NFPA 12A, *Standard on Halon 1301 Fire Extinguishing Systems*; and NFPA 17, *Standard for Dry Chemical Extinguishing Systems*.

This is the first point in the *Code* where automatic fire protection is required. Actually, ordinary sprinkler systems are not often effective in extinguishing liquid fires. They function by keeping things cool while the burning liquid burns away or is extinguished by other means. Fires in high flash point liquids can be extinguished by high-pressure water spray systems, and many liquid fires can be put out by foam-water sprinkler systems. Where automatic systems that are based on water are used for extinguishment, water sprinklers are usually desirable as a backup.

8-6.3 An approved means for prompt notification of fire to those within the plant and the public fire department available shall be provided. Where service is available, a public fire alarm box shall be located nearby if required by the authority having jurisdiction. It may be advisable to connect the plant system with the public system. NFPA 71, *Standard for the Installation, Maintenance and Use of Central Station Signaling Systems for Guard, Fire Alarm and Supervisory Service*; NFPA 72B, *Standard for the Installation, Maintenance and Use of Auxiliary Protective Signaling Systems for Fire Alarm Service*; NFPA 72A, *Standard for the Installation, Maintenance and Use of Local Protective Signaling Systems for Watchman, Fire Alarm and Supervisory Sevice*; NFPA 72C, *Standard for the Installation, Maintenance and Use of Remote Station Protective Signaling Systems*; NFPA 72D, *Standard for the Installation, Maintenance and Use of Proprietary Protective Signaling Systems for Watchman, Fire Alarm and Supervisory Service*; and NFPA 1221, *Standard for the Installation, Maintenance and Use of Public Fire Service Communications*, provide information on these subjects.

For greatest reliability, any system should be of a type that does not fail to

transmit a signal in the event of either accidental grounding of a circuit or a break in any single wire.

8-6.4 All plant fire protection facilities shall be adequately maintained and periodically inspected and tested to make sure they are always in satisfactory operating condition and they will serve their purpose in time of emergency.

8-7 Sources of Ignition.
8-7.1 General.
8-7.1.1 Precautions shall be taken to prevent the ignition of flammable vapors. Sources of ignition include but are not limited to open flames; lightning; smoking; cutting and welding; hot surfaces; frictional heat; static, electrical and mechanical sparks; spontaneous ignition, including heat-producing chemical reactions; and radiant heat.

This is the same as 5-6.1.

8-7.1.2 Class I liquids or Class II or Class III liquids at a temperature above their flashpoints (*see 1-1.3*) shall not be dispensed into metal containers unless the nozzle or fill pipe is in electrical contact with the container. This can be accomplished by maintaining metallic contact during filling, by a bond wire between them, or by other conductive path having an electrical resistance not greater than 10^6 ohms. Bonding is not required where a container is filled through a closed system, or the container is made of glass or other nonconducting material. NFPA 77, *Recommended Practice on Static Electricity*, provides information on static protection; NFPA 78, *Lightning Protection Code*, provides information on lightning protection.

This is the same as 5-6.2.

8-7.2 Maintenance and Repair.
8-7.2.1 When necessary to do maintenance work in a liquid processing area, the work shall be authorized by a responsible member of supervision.

A provision such as this is lacking in Chapter 5. The reasoning is that hazards involving chemical reactions may be present in addition to the hazards of flammability or combustibility.

8-7.2.2 Hot work, such as welding or cutting operations, use of spark-producing power tools, and chipping operations shall be permitted only under supervision of an individual in responsible charge. The individual in responsible charge shall make an inspection of the area to be sure that it is safe for the work to be done and that safe procedures will be followed for the work specified. NFPA 327, *Standard Procedures for Cleaning or Safeguarding Small Tanks and Containers*; NFPA 36, *Standard for Solvent Extraction Plants*; and NFPA 51, *Standard for the Installa-*

tion and Operation of Oxygen-Fuel Gas Systems for Welding and Cutting, provide information on such operations.

This is the same as 5-8.1 (and the comments there apply here) with the addition of the reference to NFPA 51[5] which, in turn, references NFPA 51B.[6] This latter document is important and should not be overlooked.

8-7.3 Electrical Equipment.

8-7.3.1 Section 8-7.3 shall apply to areas where Class I liquids are stored or handled or where Class II or Class III liquids are stored or handled at a temperature above their flash points (*see 1-1.3*). For areas where Class II or Class III liquids only are stored or handled at a temperature below their flash points, the electrical equipment may be installed in accordance with the provisions of NFPA 70, *National Electrical Code*, for ordinary locations; however, care shall be used in locating electrical apparatus to prevent hot metal from falling into open equipment.

This is the same as 5-7.1.

8-7.3.2 All electrical equipment and wiring shall be of a type specified by and shall be installed in accordance with NFPA 70, *National Electrical Code*.

This is the same as 5-7.2.

8-7.3.3 So far as it applies, 8-7.3.5 shall be used to delineate and classify areas for the purpose of installation of electrical equipment under normal circumstances. In the application of classified areas, a classified area shall not extend beyond an unpierced floor, wall, roof or other solid partition. The designation of classes and divisions is defined in Chapter 5, Article 500, of NFPA 70, *National Electrical Code*.

This is the same as 5-7.3, except that the reference to 8-7.3.5 leads to Table 8-2.1, which is the same as Table 5-7.3.

8-7.3.4 The area classifications listed in 8-7.3.5 are based on the premise that the installation meets the applicable requirements of this code in all respects. Should this not be the case, the authority having jurisdiction shall have the authority to classify the extent of the area.

This is the same as 5-7.4.

8-7.3.5 Extent of classified areas shall be as follows:

This is the same as 5-7.5, except for the reference to Table 8-2.1, which is the same as Table 5-7.3.

Table 8-7.3

Location	NEC Class I Division	Extent of Classified Area
Indoors equipment installed in accordance with 5-3.5.2 where flammable vapor-air mixtures may exist under normal operations.	1	Area within 5 feet of any edge of such equipment, extending in all directions.
	2	Area between 5 feet and 8 feet of any edge of such equipment, extending in all directions. Also, area up to 3 feet above floor or grade level within 5 feet to 25 feet horizontally from any edge of such equipment.*
Outdoor equipment of the type covered in 5-3.5.2 where flammable vapor-air mixtures may exist under normal operations.	1	Area within 3 feet of any edge of such equipment, extending in all directions.
	2	Area between 3 feet and 8 feet of any edge of such equipment extending in all directions. Also, area up to 3 feet above floor or grade level within 3 feet to 10 feet horizontally from any edge of such equipment.
TANK — ABOVEGROUND**		
Shell, Ends, or Roof and Dike Area	2	Within 10 feet from shell, ends or roof of tank. Area inside dikes to level of top of dike.
Vent	1	Within 5 feet of open end of vent, extending in all directions.
	2	Area between 5 feet and 10 feet from open end of vent, extending in all directions.
Floating Roof	1	Area above the roof and within the shell.

*The release of Class I liquids may generate vapors to the extent that the entire building, and possibly a zone surrounding it, should be considered a Class I, Division 2 location.

**For Tanks — Underground, see Section 7-5 of Chapter 7.

Table 8-7.3 — *continued*

Location	NEC Class I Division	Extent of Classified Area
DRUM AND CONTAINER FILLING		
Outdoors, or Indoors with Adequate Ventilation	1	Within 3 feet of vent and fill opening, extending in all directions.
	2	Area between 3 feet and 5 feet from vent or fill opening, extending in all directions. Also up to 18 inches above floor or grade level within a horizontal radius of 10 feet from vent or fill opening.
PUMPS, BLEEDERS, WITHDRAWAL FITTINGS, METERS AND SIMILAR DEVICES		
Indoors	2	Within 5 feet of any edge of such devices, extending in all directions. Also up to 3 feet above floor or grade level within 25 feet horizontally from any edge of such devices.
Outdoors	2	Within 3 feet of any edge of such devices, extending in all directions. Also up to 18 inches above grade level within 10 feet horizontally from any edge of such devices.
PITS		
Without Mechanical Ventilation	1	Entire area within pit if any part is within a Division 1 or 2 classified area.
With Mechanical Ventilation	2	Entire area within pit if any part is within a Division 1 or 2 classified area.
Containing Valves, Fittings or Piping, and Not Within a Division 1 or 2 Classified Area	2	Entire pit.
DRAINAGE DITCHES, SEPARATORS, IMPOUNDING BASINS	2	Area up to 18 inches above ditch, separator or basin. Also up to 18 inches above grade within 15 feet horizontally from any edge.

SI Units: 1 in. = 25.40 mm; 1 ft = 0.3048 m.

PROCESSING PLANTS   211

This is the same as Table 5-7.3.

8-7.3.6 Where the provisions of 8-7.3.1, 8-7.3.2, 8-7.3.3, 8-7.3.4, and 8-7.3.5 require the installation of electrical equipment suitable for Class I, Division 1 or Division 2 locations, ordinary electrical equipment including switchgear may be used if installed in a room or enclosure which is maintained under positive pressure with respect to the classified area. Ventilation make-up air shall be uncontaminated by flammable vapors. NFPA 496, *Standard for Purged Enclosures for Electrical Equipment in Hazardous Locations*, provides details for these types of installations.

This is the same as 5-7.6.

8-8 Housekeeping.

8-8.1 Maintenance and operating practices shall be in accordance with established procedures which will tend to control leakage and prevent the accidental escape of liquids. Spills shall be cleaned up promptly.

This is the same as 5-9.1.

8-8.2 Adequate aisles shall be maintained for unobstructed movement of personnel and so that fire protection equipment can be brought to bear on any part of the processing equipment.

This is the same as 5-9.2, except that only processing equipment is referred to. Obviously, storage, use, and physical operations, if any, should be treated in the same way.

8-8.3 Combustible waste material and residues in a building or operating area shall be kept to a minimum, stored in closed metal waste cans, and disposed of daily.

This is the same as 5-9.3.

8-8.4 Ground area around buildings and operating areas shall be kept free of tall grass, weeds, trash or other combustible materials.

This is the same as 5-9.4.

REFERENCES CITED BY *CODE*

(*These publications comprise a part of the requirements to the extent called for by the* **Code**.)

NFPA 10, *Portable Fire Extinguishers*, NFPA, Boston, 1978.

NFPA 11, *Foam Extinguishing Systems*, NFPA, Boston, 1978.

NFPA 12, *Carbon Dioxide Extinguishing Systems*, NFPA, Boston, 1980.

NFPA 12A, *Halon 1301 Fire Extinguishing Systems*, NFPA, Boston, 1980.

NFPA 14, *Standpipe & Hose Systems*, NFPA, Boston, 1980.

NFPA 15, *Water Spray Fixed Systems*, NFPA, Boston, 1979.

NFPA 17, *Dry Chemical Extinguishing Systems*, NFPA, Boston, 1975.

NFPA 24, *Outside Protection*, NFPA, Boston, 1977.

NFPA 327, *Cleaning or Safeguarding Small Tanks and Containers*, NFPA, Boston, 1975.

NFPA 36, *Solvent Extraction Plants*, NFPA, Boston, 1978.

NFPA 496, *Purged and Pressurized Enclosures for Electrical Equipment*, NFPA, Boston, 1974.

NFPA 51, *Oxygen-Fuel Gas Systems for Welding and Cutting*, NFPA, Boston, 1977.

NFPA 69, *Explosion Prevention Systems*, NFPA, Boston, 1978.

NFPA 70, *National Electrical Code*, NFPA, Boston, 1981.

NFPA 71, *Central Station Signaling Systems*, NFPA, Boston, 1977.

NFPA 72A, *Local Protective Signaling Systems*, NFPA, Boston, 1979.

NFPA 72B, *Auxiliary Protective Signaling Systems*, NFPA, Boston, 1979.

NFPA 72C, *Remote Station Protective Signaling Systems*, NFPA, Boston, 1975.

NFPA 72D, *Proprietary Protective Signaling Systems*, NFPA, Boston, 1979.

NFPA 78, *Lightning Protection Code*, NFPA, Boston, 1977.

NFPA 80, *Fire Doors and Windows*, NFPA, Boston, 1979.

NFPA 90A, *Air Conditioning and Ventilating Systems*, NFPA, Boston, 1978.

NFPA 91, *Blower and Exhaust Systems, Dust, Stock & Vapor Removal or Conveying*, NFPA, Boston, 1973.

NFPA 101, *Life Safety Code*, NFPA, Boston, 1981.

NFPA 1221, *Public Fire Service Communications*, NFPA, Boston, 1980.

NFPA 68, *Explosion Venting Guide*, NFPA, Boston, 1978.

NFPA 77, *Recommended Practice on Static Electricity*, NFPA, Boston, 1977.

NFPA 204, *Smoke and Heat Venting*, NFPA, Boston, 1968.

REFERENCES CITED IN COMMENTARY

[1] NFPA 101, *Life Safety Code*, NFPA, Boston, 1981.

[2] U.S. Coast Guard, *The Effectiveness of "Fire Gas Recirculation" Extinguishing Systems on Shipboard Machinery Space Fires*, National Technical Information Service, Springfield, VA 22161.

[3] NFPA 68, *Explosion Venting Guide*, NFPA, Boston, 1978.

[4] McKinnon, G.P., ed. *Fire Protection Handbook*, 14th ed., NFPA, Boston, 1976.

[5] NFPA 51, *Oxygen-Fuel Gas Systems for Welding and Cutting*, NFPA, Boston, 1977.

[6] NFPA 51B, *Fire Prevention in Use of Cutting and Welding Processes*, NFPA, Boston, 1977.

9

Refineries, Chemical Plants, and Distilleries

9-1 Storage.

9-1.1 Liquids shall be stored in tanks, in containers, or in portable tanks. Tanks shall be installed in accordance with Chapter 2 of this code.

> The omission of a reference to Chapter 4 is intentional. Paragraphs 4-1.2(f), 4-2.3.1, and 4-2.3.2 effectively exclude most distilling, storage, and handling practices concerning barreled and bottled liquids. The lack of a restriction on use of drums and other containers in refineries and chemical plants is based on considerations discussed under 8-4.3.1. Containers of liquids are often used for finished products. The provisions of Chapter 4 for storage of filled containers in warehouses should be followed, as applicable.

9-1.2 Tanks for the storage of liquids in tank farms and in locations other than process areas shall be located in accordance with 2-2.1 and 2-2.2.

> Unusual materials of construction may be found. See 3-2.4.

9-1.3 Piping, Valves and Fittings. Piping systems shall be in accordance with Chapter 3.

9-2 Wharves.

> See Section 6-4, Wharves, which reads: "The term wharf shall mean any wharf, pier, bulkhead or other structure over or contiguous to navigable water, the primary function of which is the transfer of liquid cargo in bulk between shore installations and any tank vessel, ship, barge, lighter boat or other mobile floating craft; and this section shall apply to all such installations except marine service stations as covered in Chapter 7. If liquids are handled in bulk quantities across general purpose piers or wharves **NFPA 87**, *Standard for the Construction and Protection of Piers and Wharves*,[1] shall be followed."

9-2.1 Wharves handling flammable or combustible liquids shall be in accordance with Section 6-4.

See Section 6-4.

9-3 Location or Process Units.

9-3.1 Process units shall be located so that they are accessible from at least one side for the purpose of fire control. Where topographical conditions are such that liquids can flow from a processing area so as to constitute a fire hazard to property of others, provision shall be made to divert or impound the flow by curbs, drains, or other suitable means.

This is more restrictive than is the case in Chapters 2, 5, and 8 where dikes that might retain a burning liquid are permitted on a property line, although distances to exposures are measured from the tank or vessel containing the liquid that might be involved in a spill fire. Here, the implication is that such a spill fire shall be diverted or impounded so that such a fire will not constitute an exposure hazard.

9-4 Fire Control.

9-4.1 Smoking shall be permitted only in approved areas.

In Chapters 5 and 8, smoking is mentioned as only one potential source of ignition. See 5-5.1 and 8-7.1.1. Also see the comments under 5-1.1.

9-4.2 Hot work, such as welding or cutting operations, use of spark-producing power tools, and chipping operations shall be permitted only under supervision of an individual in responsible charge. The individual in responsible charge shall make an inspection of the area to be sure that it is safe for the work to be done and that safe procedures will be followed for the work specified. NFPA 327, *Standard for Cleaning or Safeguarding Small Tanks and Containers*, and NFPA 36, *Standard for Solvent Extraction Plants*, provide information on such operations.

9-4.3 Maintenance and operating practices shall be in accordance with established procedures which will tend to control leakage and prevent the accidental escape of flammable or combustible liquids. Spills shall be cleaned up promptly.

9-4.4 Portable fire extinguishment and control equipment shall be provided in such quantities and types as are needed for the special hazards of operation and storage. NFPA 10, *Standard for Portable Fire Extinguishers*, provides information as to the suitability of various types of extinguishers.

9-4.5 Water shall be available in volume and at adequate pressure to supply water hose streams, foam producing equipment, automatic sprinklers or water

REFINERIES, CHEMICAL PLANTS, AND DISTILLERIES

spray systems as the need is indicated by the special hazards of operation and storage.

This is the same as 5-5.2, except that "dispensing" is dropped since it is not a common practice in operations covered by this chapter.

9-4.6 Special extinguishing equipment such as that utilizing foam, inert gas, or dry chemical shall be provided as the need is indicated by the special hazards of operation and storage.

This is the same as 5-5.3, except that "dispensing" is dropped.

9-4.7 An approved fire alarm system is recommended for prompt notification of fire. Where service is available, it is recommended that a public fire alarm box be located nearby. It may be advisable to connect the plant system with the public system. NFPA 72D, *Standard for the Installation, Maintenance and use of Proprietary Protective Signaling Systems for Watchman, Fire Alarm and Supervisory Service*, provides information on this subject.

9-4.8 An emergency control organization consistent with provided equipment and available personnel shall be established, and appropriate procedures specified, to cope with fire or other emergencies. Plant personnel assigned to the emergency control organization shall be trained in their duties.

9-4.9 An approved means for prompt notification of fire to those within the plant and the public fire department available shall be provided.

REFERENCE CITED BY *CODE*

(This publication comprises a part of the requirements to the extent called for by the **Code**.*)*

NFPA 10, *Portable Fire Extinguishers*, NFPA, Boston, 1978.

REFERENCE CITED IN COMMENTARY

[1]NFPA 87. *Standard for the Construction and Protection of Piers and Wharves*, NFPA, Boston, 1980.

Appendix A

Emergency Relief Venting for Fire Exposure for Aboveground Tanks

This Appendix is not part of the requirements of this NFPA document, but is included for information purposes only.

The requirements for emergency venting given in Table 2-8 and the modification factors in 2-2.5.7 are derived from a consideration of:

1. Probable maximum rate of heat transfer per unit area;

2. Size of tank and the percentage of total area likely to be exposed;

3. Time required to bring tank contents to boil;

4. Time required to heat unwet portions of the tank shell or roof to a temperature where the metal will lose strength;

5. Effect of drainage, insulation and the application of water in reducing fire exposure and heat transfer.

Table 2-8 is based on a composite curve which is considered to be composed of three straight lines when plotted on log-log paper. The curve may be defined in the following manner:

The first straight line is drawn on log-log paper between the point 400,000 Btu/hr, at 20 sq ft (1.858 m^2) exposed surface area and the point 4,000,000 Btu/hr, at 200 sq ft (18.58 m^2) exposed surface area. The equation for this portion of the curve is $Q = 20,000A$.

The second straight line is drawn on log-log graph paper between the points 4,000,000 Btu/hr, at 200 sq ft (18.58 m^2) exposed surface area and 9,950,000 Btu/hr, at 1,000 sq ft (92.9 m^2) exposed surface area. The equation for this portion of the curve is $Q = 199,300A^{.566}$.

The third straight line is plotted on log-log graph paper between the points 9,950,000 Btu/hr, at 1,000 sq ft (92.9 m^2) exposed surface area and 14,090,000 Btu/hr, at 2,800 sq ft (260.12 m^2) exposed surface area. The equation for this portion of the curve is $Q = 963,400A^{.338}$.

Q = 20,000 A		Q = 199,300 A^.566		Q = 963,400 A^.888	
A	Q	A	Q	A	Q
20	400,000	200	4,000,000	1,000	10,000,000
30	600,000	250	4,539,000	1,200	10,593,000
40	800,000	300	5,032,000	1,400	11,122,000
50	1,000,000	350	5,491,000	1,600	11,601,000
60	1,200,000	400	5,922,000	1,800	12,040,000
70	1,400,000	500	6,719,000	2,000	12,449,000
80	1,600,000	600	7,450,000	2,400	13,188,000
90	1,800,000	700	8,129,000	2,800	14,000,000
100	2,000,000	800	8,768,000	and over	
120	2,400,000	900	9,372,000		
140	2,800,000	1,000	10,000,000		
160	3,200,000				
180	3,600,000				
200	4,000,000				

For areas exceeding 2,800 sq ft (260.12 m²) it has been concluded that complete fire involvement is unlikely, and loss of metal strength from overheating will cause failure in the vapor space before development of maximum possible vapor evolution rate. Therefore, additional venting capacity beyond the vapor equivalent of 14,090,000 Btu/hr. will not be effective or required.

For tanks and storage vessels designed for pressures over 1 psig, additional venting for exposed surfaces beyond 2,800 sq ft (260.12 m²) is believed to be desirable because, under these storage conditions, liquids are stored close to their boiling points. Therefore, the time to bring the container contents to boiling conditions may not be significant. For these situations a heat input value should be determined on the basis of

$$Q - 21,000 A^{0.82}$$

The flow capacities are based on the assumption that the stored liquid will have the characteristics of hexane, and the vapor liberated has been transposed to equivalent free air at 60 °F (15.6 °C) and 14.7 psia (101.3 kPa) by using appropriate factors in:

$$CFH = \frac{70.5Q}{L\sqrt{M}}$$ where 70.5 is the factor for converting pounds of gas to cubic feet of air, Q = the

total heat input per hour expressed in Btu; L = latent heat of vaporization; and M = molecular weight.

The equations used to derive the table at the beginning of this Appendix were reached by consideration of three different approaches. Each was intended to yield an adequately conservative value which would maintain exposed tanks below a hazardous internal overpressure. Significant data were considered to be those from tests conducted by the Rubber Reserve Corporation in Baytown, Texas, in 1944, and by the American Petroleum Institute in 1947. The Rubber Reserve Corporation test boiled water in a tank exposed to a gasoline fire, and the API tests heated water using a kerosene fire. A summary of the test results and an extensive bibliography are given in the 1973 National Academy of Sciences publication, *Pressure-Relieving Systems for Marine Cargo Bulk Liq-*

uid Containers. **This can be purchased from the National Academy of Sciences, Office of Publications, 2101 Constitution Avenue, Washington, DC 20418.**

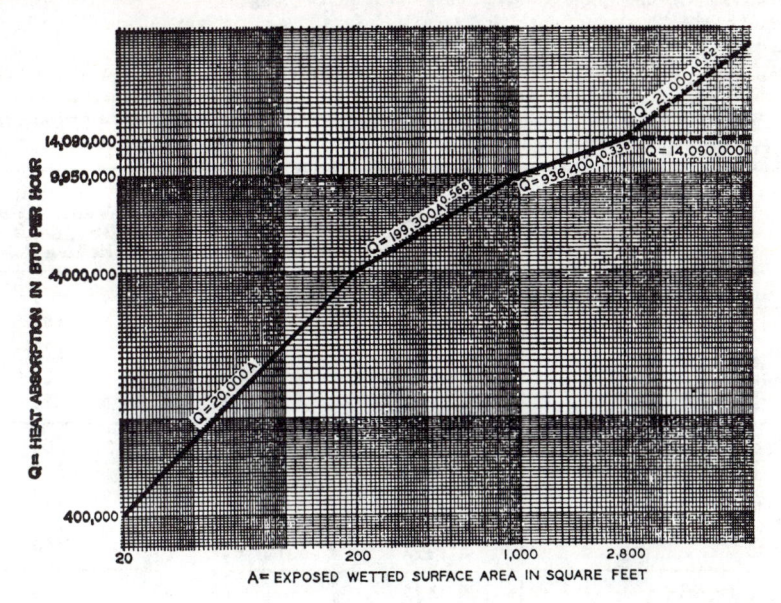

Curve for Determining Requirements for Emergency Venting During Fire Exposure.

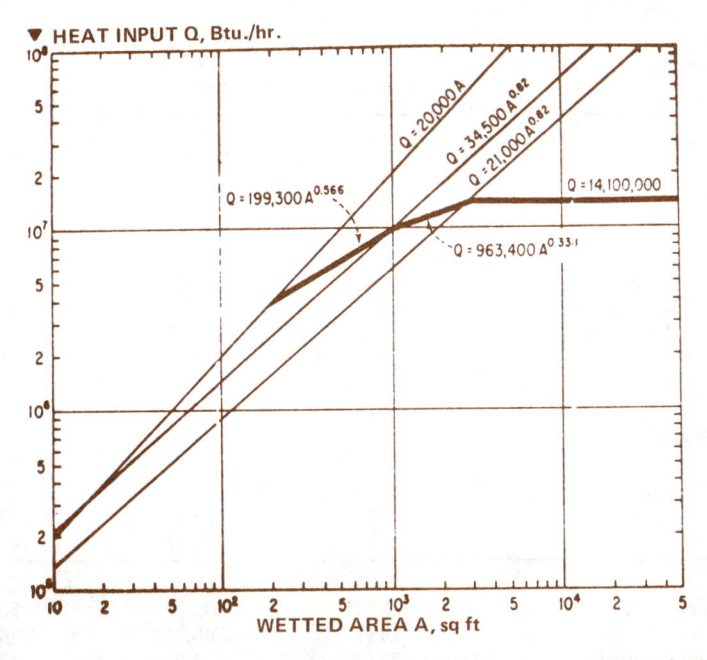

Figure A-1. Curve for determining requirements for emergency venting during fire exposure (without graph background).

No consideration has been given to possible expansion from the heating of the vapor above the boiling point of the liquid, its specific heat, or the difference in density between the discharge temperature and 60°F (15.6°C), since some of these changes are compensating.

Since tank vent valves are ordinarily rated in CFH standard air, the figures derived from Table 2-8 may be used with the appropriate tank pressure as a basis for valve selection.

The attached table gives for a variety of chemicals the constants which can be used to compute the vapor generated and equivalent free air for liquids other than hexane, where greater exactness is desired. Inspections of the table will show that the use of hexane in deriving Table 2-8 provides results which are within an acceptable degree of accuracy for the listed liquids.

Chemical	$L\sqrt{M}$	Molecular Weight	Heat of Vaporization Btu per Lb. at Boiling Point
Acetaldehyde	1673	44.05	252
Acetic acid	1350	60.05	174
Acetic anhydride	1792	102.09	177
Acetone	1708	58.08	224
Acetonitrile	2000	41.05	312
Acrylonitrile	1930	53.05	265
n-Amyl alcohol	2025	88.15	216
iso-Amyl alcohol	1990	88.15	212
Aniline	1795	93.12	186
Benzene	1493	78.11	169
n-Butyl acetate	1432	116.16	133
n-Butyl alcohol	2185	74.12	254
iso-Butyl alcohol	2135	74.12	248
Carbon disulfide	1310	76.13	150
Chlorobenzene	1422	112.56	134
Cyclohexane	1414	84.16	154
Cyclohexanol	1953	100.16	195
Cyclohexanone	1625	98.14	164
o-Dichlorobenzene	1455	147.01	120
cis-Dichloroethylene	1350	96.95	137
Diethyl amine	1403	73.14	164
Dimethyl acetamide	1997	87.12	214
Dimethyl amine	1676	45.08	250
Dimethyl formamide	2120	73.09	248
Dioxane (diethylene ether)	1665	88.10	177
Ethyl acetate	1477	88.10	157
Ethyl alcohol	2500	46.07	368
Ethyl chloride	1340	64.52	167
Ethylene dichloride	1363	98.97	137
Ethyl ether	1310	74.12	152
Furan	1362	68.07	165
Furfural	1962	96.08	200
Gasoline	1370-1470	96.0	140-150
n-Heptane	1383	100.20	138
n-Hexane	1337	86.17	144
Hydrogen cyanide	2290	27.03	430
Methyl alcohol	2680	32.04	474

Chemical	L √M	Molecular Weight	Heat of Vaporization Btu per Lb. at Boiling Point
Methyl ethyl ketone	1623	72.10	191
Methyl methacrylate	1432	100.14	143
n-Octane	1412	114.22	132
n-Pentane	1300	72.15	153
n-Propyl acetate	1468	102.13	145
n-Propyl alcohol	2295	60.09	296
iso-Propyl alcohol	2225	60.09	287
Tetrahydro furan	1428	72.10	168
Toluene	1500	92.13	156
Vinyl acetate	1532	86.09	165
o-Xylene	1538	106.16	149

NOTE: For data on other chemicals, see chemistry handbook.

APPROXIMATE WETTED AREAS FOR HORIZONTAL TANKS

(Wetted Area Equals 75 Percent Total Area)

Tank Diameter, Feet	3	4	5	6	7	8	9	10	11	12
Tank Length, Feet	APPROXIMATE WETTED AREA OF TANKS WITH FLAT HEADS									
3	32									
4	39	55								
5	46	65	88							
6	53	74	100	128						
7	60	84	112	142	173					
8	67	93	124	156	190	226				
9	74	102	136	170	206	245	286			
10	81	112	147	184	223	264	308	353		
11	88	121	159	198	239	283	329	377	428	
12	95	131	171	213	256	301	350	400	454	509
13	102	140	183	227	272	320	371	424	480	537
14	109	150	194	241	289	339	393	447	506	565
15	116	159	206	255	305	358	414	471	532	594
16	123	169	218	269	322	377	435	495	558	622
17	130	178	230	283	338	395	456	518	584	650
18	137	188	242	298	355	414	477	542	610	678
19		197	253	312	371	433	499	565	636	707
20		206	265	326	388	452	520	589	662	735
21		216	277	340	404	471	541	612	688	763

SI Units: 1 ft = 0.3048 m; 1 sq ft = 0.0929 m².

APPROXIMATE WETTED AREAS FOR HORIZONTAL TANKS — Continued

Tank Diameter, Feet	3	4	5	6	7	8	9	10	11	12
Tank Length, Feet	APPROXIMATE WETTED AREA OF TANKS WITH FLAT HEADS										
22	225	289	354	421	490	562	636	714	792
23	235	300	368	437	508	584	659	740	820
24	244	312	383	454	527	605	683	765	848
25		324	397	470	546	626	706	791	876
26		336	411	487	565	647	730	817	905
27		347	425	503	584	668	754	843	933
28		359	440	520	603	690	777	869	961
29		371	454	536	621	711	801	895	989
30		383	468	553	640	732	824	921	1018
31		395	482	569	659	753	848	947	1046
32			496	586	678	775	871	973	1074
33			510	602	697	796	895	999	1103
34			524	619	715	817	918	1025	1131
35			539	635	734	838	942	1051	1159
36			553	652	753	860	966	1077	1187
37			567	668	772	881	989	1103	1216
38				685	791	902	1013	1129	1244
39				701	810	923	1036	1155	1272
40				718	828	944	1060	1181	1301
41				734	847	966	1083	1207	1329
42				751	866	987	1107	1233	1357
43				767	885	1008	1130	1259	1385
44					904	1029	1154	1284	1414
45					923	1051	1178	1310	1442
46					941	1072	1201	1336	1470
47					960	1093	1225	1362	1498
48					979	1114	1248	1388	1527
49					998	1135	1272	1414	1555
50						1157	1295	1440	1583
51						1178	1319	1466	1612
52						1199	1342	1492	1640
53						1220	1366	1518	1668
54						1246	1389	1544	1696
55						1263	1413	1570	1725
56							1437	1593	1753
57							1460	1622	1781
58							1484	1648	1809
59							1507	1674	1839

SI Units: 1 ft = 0.3048 m; 1 sq ft = 0.0929 m².

APPROXIMATE WETTED AREAS FOR HORIZONTAL TANKS — Continued

Tank Diameter, Feet	3	4	5	6	7	8	9	10	11	12
Tank Length, Feet	\multicolumn{10}{c}{APPROXIMATE WETTED AREA OF TANKS WITH FLAT HEADS}
60								1531	1700	1866
61									1726	1894
62									1752	1923
63									1778	1951
64									1803	1979
65									1829	2007
66									1855	2036
67										2064
68										2092
69										2120
70										2149
71										2177
72										2205

SI Units: 1 ft = 0.3048 m; 1 sq ft = 0.0929 m^2.

Significant interpretations of data from prior tests and from fire experience were also available in the form of equations. These took the form of straight lines on log-log paper: one was $Q = 20,000A$, where Q was Btu/hr and A was sq ft of wetted surface exposed to fire. This was based on work reported by Duggan, Gilmour, and Fisher in a paper in the NFPA *Quarterly*, Vol. 37, October, 1943. Similar formulas in use were $Q = 21,000A^{0.82}$, used by the American Petroleum Institute prior to 1966, and $Q = 34,500^{0.82}$, still in use in NFPA 58, *Standard on the Handling and Storage of Liquefied Petroleum Gases*. In NFPA 58, A is the total surface area of the tank. It is obvious that it is easier to engulf a small tank in flame than it is to engulf a large tank. Work done by West Virginia University for Union Carbide Corporation showed that a sustained external temperature of about 1,600 °F (871 °C) would be required to bring most flammable liquids [except those belonging to Class IA which boil below 100 °F (37.8 °C)] to a boiling point in about 15 minutes. However, at that external temperature, the metal above the wetted portion would soften and fail in less than 10 minutes; thus, the tank would be self-venting and emergency vents would not be needed. Accordingly, $Q = 20,000A$ was used for tanks having A equal to or less than 200 sq ft (18.5 sq m). This works out to a tank having a capacity of up to about 2,500 gal (9462.5 L). It gives a point on the chart at $A = 200$, $Q = 4,000,000$. The line for the equation $Q = 34,500A^{0.82}$ intersects the 1,000 sq ft (92.9 m^2) abscissa at the point $A = 1,000$, $Q = 9,950,000$. $A = 1,000$ was selected on the basis of surface area. It is equivalent to about a 25,000 to 30,000 gal tank. The line for the equation $Q =$

21,000A$^{0.82}$ gives the point A = 2,800, Q = 14,090,000. The point at A = 2,800 was selected on the basis of being close to Q = 14,090,000. This value of Q is equivalent to about 743,000 cu ft (21 042 m³) of air, using the formula

$$\text{CFH} = \frac{70.5Q}{L\sqrt{M}}$$ where $L\sqrt{M}$ is 144, the figure for n-Hexane (CH$_3$ CH$_2$ CH$_2$ CH$_2$ CH$_2$ CH$_3$). This is the amount of vapor that would be released by an 18 in. (45.72 cm) emergency vent. It is felt by the Technical Committee that the time required to boil a liquid requiring a vent larger than 18 in. (45.72 cm) would soften the unwetted metal so that the tank would be self-venting. However, when large low-pressure tanks are used, it is probable that the contents would be low boiling. For such a case, the line Q = 21,000A$^{0.82}$ is extended.

The formula cu ft/hr of free air at 60 °F (15.6 °C) = $\frac{70.5Q}{L\sqrt{M}}$ is derived by noting that $\frac{Q}{L}$ is the amount in lbs that will be vaporized when Q is in Btu and L is the heat needed to vaporize 1 lb of the material being considered. One molecular weight, of any material, expressed in lbs, produces 359 cu ft (10 m³) of vapor at 32 °F (0 °C) and normal atmospheric pressure. If air is assumed to have a molecular weight of 28.97, $\frac{359}{\sqrt{28.97}}$ = 66.69. This figure must be multiplied by $\frac{288.72}{273.16}$ or 1.057, the ratio of the absolute temperatures in degrees Kelvin; 273.16 is 0 °C and 288.72 is 60 °F. This gives the 70.5 in the formula.

Appendix B

Abandonment or Removal of Underground Tanks

This Appendix is not part of the requirements of this NFPA document, but is included for information purposes only.

B-1 Introduction.

B-1-1 Care is required not only in the handling and use of flammable or combustible liquids but also in abandoning tanks which have held flammable or combustible liquids. This is particularly true of underground service station tanks which are most frequently used for the storage of motor fuel and occasionally for the storage of other flammable or combustible liquids, such as crankcase drainings (which may contain some gasoline). Through carelessness, explosions have occurred because flammable or combustible liquid tanks had not been properly conditioned before being abandoned.

Although accidents from these causes may be infrequent, the results can be tragic. One such case caused the death of three fire fighters and two service station employees, when vapors from an abandoned gasoline tank exploded. The vapors had accumulated in the crawl space under the service station building. Therefore, it is certainly important enough to recommend practices that would control the hazards associated with abandoned underground tanks. The intent is to avoid such tanks creating a hazard at some indefinite time in the future.

B-1-2 In order to prevent accidents caused by improper conditioning, it is recommended that the procedures outlined below be followed when underground tanks are removed, abandoned or temporarily taken out of service.

B-1-3 Underground tanks taken out of service may be safeguarded or disposed of by any one of the three following means:

(a) Placed in a "temporarily out of service" condition. Tanks should be rendered "temporarily out of service" only when it is planned that they will be returned to active service within a reasonable period or pending removal or abandonment within 90 days.

(b) Abandoned in place, with proper safeguarding.

(c) Removed.

A decision as to which of the preceding methods should be followed is based on negotiations between the property owner and the authority having jurisdiction, consideration of the previous use of the tank, and the intended use of the property.

B-1-4 In cases where tanks are either rendered "temporarily out of service" or permanently abandoned, records should be kept of tank size, location, date of abandonment, and method used for placing the abandoned tank in a safe condition.

B-1-5 Procedures for carrying out each of the above methods of disposing of underground tanks are described in the following sections. No cutting torch or other flame or spark producing equipment shall be used until the tank has been completely purged or otherwise rendered safe. In each case, the numbered steps given shall be carried out successively.

B-2 Rendering Tanks "Temporarily Out of Service."

B-2-1 Cap or plug all lines such as fill line, gage opening, pump suction, and vapor return. Secure against tampering.

B-2-2 Disconnect piping at all tank openings.

B-3 Abandoning Underground Tanks in Place.

B-3-1 Remove all flammable or combustible liquid from the tank and from all connecting lines.

B-3-2 Disconnect the suction, inlet, gage, and vent lines.

B-3-3 Fill the tank completely with an inert solid material. Cap remaining undergound piping.

B-4 Removal of Underground Tanks.

B-4-1 Remove all flammable or combustible liquids from tank and from connecting lines.

B-4-2 Disconnect piping at all tank openings. Remove sections of connecting lines which are not to be used further and cap or plug all tank openings. After removal, the tank may be gas freed on the premises if it can be done safely at that location, or may be transported to an area not accessible to the public and the gas freeing completed at that location.

B-5 Disposal of Tanks.

B-5-1 If a tank is to be disposed of as junk, it should be retested for flammable vapors, and, if necessary, rendered gas free. After junking and before releasing to junk dealer, a sufficient number of holes or openings should be made in it to render it unfit for further use. NFPA 327, *Standard Procedures for Cleaning or Safeguarding Small Tanks and Containers*, provides information on safe procedures for such operations.

> The reason for making holes in the tank is to discourage possible future use of it as a container for some edible products which would be contaminated by residual deposits if the tank had ever been used for gasoline containing lead or other toxic additives.

B-6 Reuse of Underground Tanks.

B-6-1 Used tanks which are to be reused for flammable or combustible liquid service should meet all the requirements of this code for the installation of underground tanks.

Appendix C

This Appendix is not part of the requirements of this NFPA document, but is included for information purposes only.

The following contains additional information and recommendations bearing the same number as the text of the *Flammable and Combustible Liquids Code* to which they apply:

C-4-4 The preferred method of storage of liquids in buildings is in cutoff rooms or in attached buildings rather than in inside rooms because of fire department accessibility and the advantages of providing explosion venting where needed.

C-4-6.2 (a) Sprinkler system densities and areas of application presented in this appendix are based upon limited test data and fire experience. Design criteria in this appendix do not apply to storage in plastic drums. (*See Appendix D for additional information on this subject.*)

(b) For design criteria for specific installations, insurance engineers, fire protection consultants, and other knowledgeable persons should be consulted.

(c) **Palletized and Solid Pile Storage.** For protected storage of liquids, as specified in Table 4-6.1(a), automatic sprinkler protection should be provided in accordance with Table C-4-6.2(a).

(d) **Rack Storage.** In protected storage of liquids arranged, as specified in Table 4-6.1(b), automatic sprinkler protection should be provided in accordance with Tables C-4-6.2(b) and C-4-6.2(c), as applicable, except that racks with solid shelves should be provided with in-rack sprinklers at every tier or level.

> The high temperature and ordinary temperature headings in Table C-4-6.2(a) refer to the temperature ratings of the sprinkler heads. Ordinary temperature head ratings are rated at 135°F (57.8°C) to 170°F (76.7°C). For purposes of this *Code*, high temperature rating heads are those rated above 170°F (76.7°C). The delay in operation of the higher temperature sprinkler head tends to limit sprinkler operation to areas above or near the fire area so fewer heads will operate and the chance of overtaxing the system is diminished. [See Table C-4-6.2(a)]. Overtaxing a sprinkler system results from too many heads opening, and also from changes in discharge patterns due to lower pressure. For example, using 285°F (140.9°C) sprinkler heads in lieu of 160°F (71.1°C) heads will cause fewer total heads to fuse in a fire situation and thus keep the individual sprinkler head nozzle pressure higher. As more heads fuse, the total water flow is divided between the heads and the reduced flow per head causes a lower nozzle pressure. Note that if higher temperature sprinkler heads are used, the maximum area over which sprinklers may be expected to operate is reduced.

The effect of the storage arrangement is also considered. **Higher piles of storage may have a tendency to shield the fire from effective application of the sprinkler discharge. Pile storage also allows for the possibility of container rupture due to collapsing piles and falling containers. Portable tanks can be stored in greater quantities per pile because of emergency relief vents on portable tanks. However, the potential impingement on other storage of flames from emergency relief vents must be considered in the storage design.**

C-4-6.2.1 (a) Automatic aqueous film forming foam (AFFF)-water sprinkler systems for container storage of liquids has been shown to be an acceptable method for providing fixed protection. (*See Appendix D for additional information on this subject.*)

(b) For design criteria for specific installations, insurance engineers, fire protection consultants and other knowledgeable persons should be consulted.

(c) Rack storage of liquids in containers [drums of 55 gal (208.1 L) capacity] stored on-end on wood pallets on conventional double-row racks to a maximum height of storage of 25 ft (7.62 m) should be provided protection in accordance with Table C-4-6.2.1.

Automatic Sprinkler Protection for Solid Pile and Palletized Storage of Liquids in Containers and Portable Tanks.

Notes to Table C-4-6.2(c) *(See page 235.)*

(1) See Table 4-6.1(b) and C-4-6.2(b) for additional information pertaining to protected rack storage.

(2) Additional in-rack protection required for solid shelves, as indicated in C-4-6.2(d).

(3) See 4-6.3 for types of racks permitted.

(4) See 4-6.5 for additional information pertaining to in-rack sprinklers.

(5) Minimum hose stream demand includes small hand hose (1½ in.) required in 4-7.1.3.

(6) The design area contemplates the use of wet pipe systems. Where dry pipe systems are required, it introduces a possible delay which needs to be compensated for by increased areas of application (plus 30 percent).

APPENDIX C   231

Table C-4-6.2(a) Automatic Sprinkler Protection for Solid Pile and Palletized Storage of Liquids in Containers and Portable Tanks

Storage Conditions		Ceiling Sprinkler Design and Demand						
			Area (sq ft)					
Class Liquid	Container Size and Arrangement	Density gpm/sq ft	High Temp.	Ord. Temp.	Maximum Spacing	Minimum Hose Stream Demand (gpm)	Minimum Duration Sprinklers & Hose Streams
IA	5 gal. or less, with/without cartons, palletized or solid pile	0.30	3000	5000	100 sq ft	750	2 hrs
	flammable aerosols in cartons, palletized or solid pile	0.30	6000	10,000	100 sq ft	1000	2 hrs
	containers greater than 5 gal., on end or side, palletized or solid pile	0.60	5000	8000	80 sq ft	750	
IB, IC, & II	5 gal. or less, with/without cartons, palletized or solid pile	0.30	3000	5000	100 sq ft	500	2 hrs
	containers greater than 5 gal., on pallets or solid pile, one high	0.25	5000	8000	100 sq ft		
II	containers greater than 5 gal., on pallets or solid pile, more than one high on end or side	0.60	5000	8000	80 sq ft	750	2 hrs
IB, IC, II	portable tanks, one high	0.30	3000	5000	100 sq ft	500	2 hrs

Table C-4-6.2(a) Continued

Class Liquid	Storage Conditions — Container Size and Arrangement	Density gpm/sq ft	Area (sq ft) High Temp.	Area (sq ft) Ord. Temp.	Maximum Spacing	Minimum Hose Stream Demand (gpm)	Minimum Duration Sprinklers & Hose Streams
II	portable tanks, two high	0.60	5000	8000	80 sq ft	750	2 hrs
	5 gal. or less, with/without cartons, palletized or solid pile	0.25	3000	5000	120 sq ft	500	1 hr
	container greater than 5 gal., on pallets or solid pile, on end or sides, up to three high	0.25	3000	5000	120 sq ft	500	1 hr
III	container greater than 5 gal., on pallets or solid pile, on end or sides up to 18 feet high	0.35	3000	5000	100 sq ft	750	2 hrs
	portable tanks, one high	0.25	3000	5000	120 sq ft	500	1 hr
	portable tanks, two high	0.50	3000	5000	80 sq ft	750	2 hrs

Notes: (1) See Table 4-6.1(a) and Section 4-6 for additional information pertaining to protected palletized or solid piling of liquids.

(2) Minimum hose stream demand includes small hand hose (1½ inches) required in 4-7.1.3.

(3) The design area contemplates the use of wet pipe systems. Where dry pipe systems are required, it introduces a possible delay which needs to be compensated for by increased areas of application (plus 30 percent).

SI Units: 1 gal = 3.785 L; 1 sq ft = 0.0929 m²; 1 ft = 0.3048 m.

Table C-4-6.2(b) Automatic Sprinkler Protection Requirements for Rack Storage of Liquids in Containers of Five Gallon Capacity or Less,* in Cartons on Conventional Wood Pallets or Without Cartons but Strapped to Pallets

*Flammable Aerosols Not Included

Class Liquid	Ceiling Sprinkler Design & Demand			In-Rack Sprinkler Arrangement and Demand				Minim. Hose Stream Demand (gpm)	Minim. Duration Sprinkler & Hose Stream
	Density gpm/ sq ft	Area (sq ft) High Temp. / Ord. Temp.		Max. Spacing	Racks up to 9 ft. (2.7m) deep	Racks over 9 ft. (2.7m) to 12 ft. (3.7m) deep	Minim. Nozzle Pressure	Number of Sprinklers Operating		
I (max. 25' height)	0.40	3000	5000	80 sq ft/hd.	a) ord. temp. sprinklers 8 feet apart horizontally b) one line sprinklers above each level of storage c) locate in longitudinal flue space, staggered vertical d) shields req'd where multilevel	a) ord. temp. sprinklers 8 feet apart horizontally b) two lines sprinklers above each level of storage c) locate in transverse flue spaces, staggered vertical and within 20 in. of aisle d) shields required where multilevel	30 psi.	a) 8 sprinklers if only one level b) 6 sprinklers ea. on two levels if only two levels c) 6 sprinklers ea. on top 3 levels, if 3 or more levels d) hydraulically most remote	750	2 hrs

FLAMMABLE AND COMBUSTIBLE LIQUIDS CODE HANDBOOK

Table C-4-6.2(b) Continued

Class Liquid	Ceiling Sprinkler Design & Demand				In-Rack Sprinkler Arrangement and Demand				Minim. Hose Stream Demand (gpm)	Minim. Duration Sprinkler & Hose Stream
	Density gpm/sq ft	Area (sq ft) High Temp.	Area (sq ft) Ord. Temp.	Max. Spacing	Racks up to 9 ft. (2.7m) deep	Racks over 9 ft. (2.7m) to 12 ft. (3.7m) deep	Minim. Nozzle Pressure	Number of Sprinklers Operating		
II (max. 25' height)	0.30	3000	5000	100 sq ft/hd.	a) ord. temp. sprinklers 8 feet apart horizontally b) one line sprinklers betw. levels at nearest 10 foot vertical intervals c) locate in longitudinal flue space, staggered vertical d) shields required where multilevel	a) ord. temp. sprinklers 8 feet apart horizontally b) two lines betw. levels at nearest 10 foot vertical intervals c) locate in transverse flue spaces, staggered vertical and within 20 in. of aisle d) shields required where multilevel	30 psi.	a) hydraulically most remote — 6 sprinklers at each level, up to max. of three levels	750	2 hrs
III max.	0.25	3000	5000	120 sq ft/hd.	Same as Class II	Same as Class II	30 psi.	Same as Class II	500	2 hrs

Notes: (1) See Table 4-6.1(b) and Section 4-6 for additional information pertaining to protected rack storage.
(2) Additional in-rack protection required for solid shelves, as indicated in D-4.6.2(d).
(3) See 4-6.3 for types of racks permitted.
(4) See 4-6.5 for additional information pertaining to in-rack sprinklers.
(5) Minimum hose stream demand includes small hand hose (1½ inches) required in 4-7.1.3.
(6) The design area contemplates the use of wet pipe systems. Where dry pipe systems are required, it introduces a possible delay which needs to be compensated for by increased areas of application (plus 30 percent).

SI Units: 1 gal = 3.785 L; 1 sq ft = 0.0929 m²; 1 ft = 0.3048 m; 1 in. = 25.40 mm.

APPENDIX C   235

Table C-4-6.2(c) Automatic Sprinkler Protection for Rack Storage of Liquids in Containers Greater Than 5 Gallon Capacity

(See notes on page 230.)

Class Liquid	Ceiling Sprinkler Design & Demand					In-Rack Sprinkler Arrangement and Demand				Minim. Hose Stream Demand (gpm)	Minim. Duration Sprinkler & Hose Stream
	Density gpm/sq ft	Area (sq ft) High Temp.	Ord. Temp.	Max. Spacing		On-Side Storage Racks up to 9 ft.	On-End Storage (on pallets) up to 9 ft. deep racks	Minim. Nozzle Pressure	Number of Sprinklers Operating		
IA (max. 25' height)	0.60	3000	5000	80 sq ft/hd.		a) ord. temp. sprinklers 8 feet apart horizontally b) one line sprinklers above each tier of storage c) locate in longitudinal flue space, staggered vertical d) shields required where multilevel	a) ord. temp. sprinklers 8 feet apart horizontally b) one line sprinklers above each tier of storage c) locate in longitudinal flue space, staggered vertical d) shields required where multilevel	30 psi.	a) hydraulically most remote 6 sprinklers at each level	1000	2 hrs
IB, IC & II (max. 25' height)	0.60	3000	5000	100 sq ft/hd.		a) see a) above b) one line sprinklers every three tiers of storage c) see c) above d) see d) above	a) see a) above b) see b) above c) see c) above d) see d) above	30 psi.	a) see a) above	750	2 hrs
III (max. 40' height)	0.25	3000	5000	120 sq ft/hd.		a) see a) above b) one line sprinklers every sixth level (maximum) c) see c) above d) see d) above	a) see a) above b) one line sprinklers every third level (maximum) c) see c) above d) see d) above	15 psi.	a) see a) above	500	1 hr

(See over.)

SI Units: 1 gal = 3.785 L; 1 sq ft = 0.0929 m²; 1 ft = 0.3048 m; 1 in. = 25.40 mm.

Table C-4-6.2.1 Automatic AFFF-Water Protection (1) Requirements for Rack Storage of Liquids* in Containers

*Flammable Aerosols Not Included

Class Liquid	Ceiling Sprinklers Design & Demand Density gpm/sq ft	Area (sq ft) High Temp.	Area (sq ft) Ord. Temp.	In-Rack Sprinkler Arrangement and Demand (4) On-End Storage, of drums (on pallets) up to 25 ft	Minimum Nozzle Pressure	Number of Sprinklers Operating	Hose Stream Demand (3)	Duration AFFF Supply	Duration Water Supply
IA, IB, IC, II	0.30	1500	2550	a) ord. temp. sprinkler up to 10 feet apart horizontally b) one line sprinklers above each level of storage c) locate in longitudinal flue space, staggered vertically d) Shields required for multilevel	30 psi.	3 sprinklers per level	500	15 min	2 hrs

Notes: (1) System shall be a closed head wet system with approved devices for proportioning AFFF.
(2) Except as modified herein, in-rack sprinklers shall be installed in accordance with *Rack Storage of Materials*, NFPA 231C.
(3) Hose stream demand includes inside hand hose (1½ inches) required in 4-7.1.3.
(4) Maximum height of storage should be limited to 25 feet.

SI Units: 1 gal = 3.785 L; 1 sq ft = 0.0929 m²; 1 ft = 0.3048 m; 1 in. = 25.40 mm.

Appendix D

This Appendix is not a part of the requirements of this NFPA document, but is included for information purposes only. For SI Units, refer to the chart at the end of this Appendix.

Appendix D explains fire test data and loss experience that were used to help promulgate protection tables that are presented in Appendix C. While these data are limited, they do illustrate the seriousness of a potential drum rupture in a fire and the primary failure mode of built-up internal pressure in combination with the weakening of the rim joint, due to localized overheating. The possibility of a BLEVE-type explosion (Boiling Liquid Expanding Vapor Explosion) is also demonstrated. Due to the many unknowns, conservative practice would be to limit all Class I liquids stored in drums to not over one drum high, since protection tables were developed with this philosophy.

Very limited fire tests and fire experience, relative to flammable aerosols, indicate the serious problem they present to the fire protection engineer. Exploding pressurized aerosol cans are to be expected, together with the flaming fire ball and rocketing action, spreading fire to a potentially larger area. The protection philosophy expressed is primarily to limit storage heights and to contemplate a larger area of application. Use of pressure-relieving can designs would be expected to affect favorably the design considerations for fixed protection.

The following tests substantiate that drums containing a Class I flammable liquid, with sprinkler protection, can rupture under severe fire exposure conditions within 2 minutes after being exposed to fire. Burning of petroleum liquid from one 55-gal drum can release well over 7,000,000 Btu. This intense heat can easily overtax a sprinkler system, unless high temperature heads are used, and can immediately involve other containers in the storage area. It is therefore important to design sprinkler systems and arrange storage so that cooling water can be applied to all exposed parts of a drum. This requires early application of a moderate volume of water or later application of larger volumes of water to effect sufficient cooling. The standard sprinkler discharges much of its water as a fine spray which can be carried away by fire gases with little cooling effect on containers or portable tanks. Special design sprinkler heads may be needed.

Recent fire incidents have further verified the need to tightly control storage in aisles and to adhere to storage requirements relative to quantities per pile, maximum pile height, and pile separation by aisles. Modifications of storage arrangements and changes in stored commodities without a reevaluation and/or modification of the fire protection design may lead to failure of the protection system with disastrous fire loss.

D-4-6(A) Fire Tests — Drum Storage:

(1) 1949 Fire Tests. A series of fire tests were made in 1949 at the Factory Mutual test center in Norwood, Massachusetts. The tests were conducted in the 15 ft-high section of the fire test building used at that time. The tests used ICC Specification 5 drums, which were 14 gage compared with the 16 gage Specification 17C drums and 18 gage Specification 17E drums used more commonly today.

The tests involved storage horizontally on metal racks up to four drums high, and palletized upright, three drums high. Test drums contained either water, gasoline, or benzene, located in the first or second tier and equipped with pressure and temperature sensing connections. The gasoline and benzene drums were piped to manual vents so that pressure could be relieved before the drums ruptured. Other drums in the array contained water or were empty.

Sprinkler protection consisted of open, old-type sprinklers, which could be manually turned on, either at the start of the fire (short preburn) or at a time simulating the first sprinkler operation (long preburn). Sprinklers were spaced either at 100 sq ft/head with a flow rate of 0.22 or 0.28 gpm/sq ft or spaced at 50 sq ft/head with a flow rate of 0.44 or 0.56 gpm/sq ft.

Gasoline was pumped through piping to designated discharge points in or near the pile at flow rates from 1 gpm to 15 gpm. In some tests, 5 or 10 gal of fuel were poured on the floor below the drums and ignited. Duration of flows were the length of time required to empty a single drum at the rate of flow used.

When sprinkler discharge was turned on immediately, the pressure that developed in the test drums was due almost entirely to the vapor pressure as the body of liquid increased in temperature. When sprinkler discharge was started, simulating normal sprinkler operation, there was a rapid pressure increase due to heating of the vapor space. This usually dropped when cooling by sprinkler discharge started.

Early tests showed that 100 sq ft spacing of sprinklers and densities of 0.22 and 0.28 gpm/sq ft would not prevent excessive temperature and pressure increases in drums. Spacing of 50 sq ft per sprinkler was used in subsequent tests. Test measurement and visual observation indicated that 0.56 gpm/sq ft provided considerably better cooling and flushing away of fuel than the 0.44 gpm/sq ft sprinkler density.

When fuel was discharged on the floor, only the bottom tier of storage was severely exposed. When fuel was discharged at a higher level, simulating a leaking drum, those drums in the immediate vicinity in upper tiers were severely exposed.

The rate of fuel flow had very little effect on the heating of any particular drum. The lower rates, 1-2 gpm, had a much longer duration and resulting exposure was greater before the 55-gal duration supply was used up.

With on-side drum storage in racks, the rate of temperature rise in the test drum on the lowest tier was 3 to 5 times as high with storage more than one drum high than it was with one-high storage. Tests with on-end palletized storage were only conducted three-high.

When 5 or 10 gal of gasoline were spilled on the floor and then ignited, the 5-gal spill gave a more severe exposure to drums because of the longer time before sprinklers would have operated. The 10-gal spill exposed more drums, but the exposure to any one drum was no more severe.

A very small leak from a drum filled with gasoline gave a very severe exposure, because of the localized exposure to the leaking drum and insufficient heat at the ceiling to operate the sprinklers.

Drums containing benzene heated much more rapidly than drums containing water because of the lower specific heat of benzene. Early pressure build-up in the vapor space is more pronounced with water, possibly because of more film vaporization on the early stages of the fire.

(2) 1967 Fire Tests. A series of fire tests were made to compare the effects of severe fire exposure to water- and heptane-filled drums. The tests were carried out in the Factory Mutual explosion tunnel, using new ICC-17E (18 gage) 55-gal drums.

A single drum was encircled with a ring of oil burners. Temperatures were measured at various points in the drum. The fuel rate to the oil burners was about 1 gpm. There was no cooling applied to the drum.

Using heptane, the drum ruptured at about 17 psig, at a drum rim temperature of 1190°F (643.4°C). The cover seam unrolled and a BLEVE-type explosion resulted, after a fire exposure of 3 to 4 minutes.

On similar tests using water, failure occurred at 40 psig after 10 minutes.

The tests indicated that the heptane-filled drum will rupture much sooner and at a much lower internal pressure than a water-filled drum. This is attributed to the fact that drums were found to leak around the joint of the rim before the rupture. The small leakage of heptane vapor through the rim joint causes a localized flame at this already weakened location on the rim, whereas steam issuing from a similar leak in a water-filled drum tends to cool the metal at this point.

(3) 1974 Fire Tests. A series of fire tests were made to evaluate protection of on-end drum storage with AFFF foam discharging from a standard sprinkler system. The tests were conducted in the 30-foot high area of the Factory Mutual test center in Rhode Island.

Based on the 1967 tests, a standard for success was that no drum should exceed 15 psig pressure.

Tests were made with water-filled drums, palletized, 2, 3, and 4 pallets high, and on racks, 5 tiers high.

Fuel was heptane, piped to the base of the top tier of storage, with a 10-gal floor spill in each case. Sprinklers were automatic, 286°F (141.1°C) heads.

Test 1: In this test, storage was 4 pallet loads high. Fuel discharge rate was 2 gpm. Sprinkler discharge density was 0.30 gpm/sq ft. The first sprinkler opened at 34 sec. Only 4 sprinklers operated, but the three-dimensional fire in the pile continued strong. Several drums bulged, 2 ruptured, and 6 exceeded 15 psig pressure.

Test 2: In this test, storage was 3 tiers high, sprinkler density was 0.60 gpm/sq ft. Other conditions were the same as Test 1.

Two sprinklers opened at about 1 minute 20 sec. A considerable number of drums were deformed. Four of the 8 monitored drums exceeded 15 psig pressure.

Test 3: This test was rack storage with 160 °F (71.1 °C) automatic sprinklers in each tier except the bottom. Fuel rate was 2 gpm. Ceiling protection was 0.30 gpm/sq ft.

Five in-rack sprinklers and 1 ceiling sprinkler opened. One drum in the first tier, which had no in-rack sprinklers, reached a pressure of 16 psig. Two drums fell from the fifth tier, due to burning away of a pallet.

Test 4: Test 4 was a repeat of Test 3, except the fuel flow rate was 15 gpm.

Eight ceiling sprinklers and 5 in-rack sprinklers operated. Ceiling temperatures reached 1665 °F (909.5 °C). One monitored drum in the first tier reached 20 psig. Several drums were bulged.

Test 5: Test 5 was a repeat of Test 2, except storage was 2 tiers high.

The fuel was a greater distance from the ceiling so sprinklers did not operate until 3½ to 4 minutes after ignition. Damage to drums was severe, with many rupturing and all eight monitored drums going over 15 psig.

Generally, results were good in rack storage, where in-rack sprinklers were provided at each tier. For palletized storage, the AFFF protection controlled the floor fire, although pallets hindered spread of foam. Ceiling sprinklers only did not adequately protect palletized storage where an elevated spill resulted in a three-dimensional fire within the pile.

Most of the ruptured drums failed at the top chime, but 1 drum developed a slow leak at a bottom chime. In Test 5, several drums were heated by a localized fire which did not open sprinklers at the roof. This slow overpressurization can lead to superheated liquid release and a resulting severe BLEVE when the drum eventually ruptures.

D-4-6(B) Fire Tests — Small Containers.

(1) 1957 Fire Test (Nonpressurized Smaller Containers). A fire test was made on 10½ ft high storage of paint in 1-gal cans in cartons. The storage was palletized, but the pallets were fire-stopped, so it was equivalent to solid piled storage. The paint varied in flashpoint from 105 °F to 170 °F (40.5 °C to 76.7 °C) (Class II and IIIA). Sprinkler protection was 160 °F (71.1 °C) heads, 10 x 10 ft, with a density of 0.23 gpm/sq ft. Ceiling height was 15 ft.

Six sprinklers operated and controlled the fire. Temperatures over the fire reached a maximum of 1100 °F (593.3 °C) and dropped below 500 °F (260 °C) after 10 minutes. Five hundred and three cans had their covers blown off and 20 cans had burst seams. The paint released from the cans was slight, but it would be much more significant if a pile had toppled over or if cans had not all been stored cover side up.

(2) 1970 Fire Test (Pressurized Containers). A fire test was made in the 30-ft high section of the Factory Mutual Rhode Island test facility. The storage was 13 and 16-oz cans of lacquer in shipping cartons stored 2 pallet by 2 pallet by 2 pallet high on racks. Storage height was 9 ft 9 in. Protection was by twelve 160°F (71.1°C) sprinklers spaced 10 ft x 10 ft providing a discharge density of 0.30 gpm/sq ft.

Fifty seconds after ignition, containers began to burst. At 62 sec, 3 sprinklers operated. The fire became more and more intense and with all 12 sprinklers operating, there was no suppressing effect. The discharge was increased to 0.50 gpm/sq ft. without effect. After about 5 minutes, the fuel was nearly exhausted. Containers were thrown to every corner of the test building.

Temperatures over the fire were over 1000°F (537.8°C) for 3½ minutes and over 1700°F (926.6°C) for 2 minutes.

D-4-6(C) Fire experience examples involving flammable and combustible liquids in containers stored in buildings.

(1) 1951 Fire. Drums of petroleum naphtha were stored temporarily in a general purpose warehouse used mainly for storing can ends in wood boxes. Storage was 1 drum high on pallets.

Two drums had small punctures and leaks near the bottom, caused either maliciously or by moving equipment. The leak was ignited, and one drum ruptured at the bottom seam. A drum rupture resulted which opened 272 sprinklers. The fire department was called promptly and they and sprinklers were able to contain the fire, helped by the low combustible concentration in the warehouse and by failure of any other drums to rupture.

Forty-two million can ends were wet down, but fire damage was limited. No explosion damage was reported. (The intensity of the BLEVE may have been limited by much of the liquid leaking from the drum before it ruptured.) Total damage was about $200,000.

(2) 1965 Fire. Pressurized containers of paint were stored 15 ft high on racks. A fire started in the top tier from a gas-fired radiant heater. Bursting containers spread burning paint over a large area, opening one hundred eighty-eight 165°F (73.9°C) sprinklers. The fire spread 25 ft along a rack but was slowed by aisles and inert material. A portion of the roof over the fire area collapsed.

(3) 1966 Fire. Pressurized containers of alcohol base hair spray and deodorant were stored palletized, 17 ft high. The fire was contained within a 1,200 sq ft pile by 107 operating sprinklers. Damage exceeded $400,000.

(4) 1971 Distribution Warehouse Fire. A sprinklered 67,000 sq ft, one-story, noncombustible warehouse for automotive equipment and supplies was destroyed by fire from undetermined cause. Storage consisted of various metal, plastic and rubber parts in cardboard cartons, plus flammable and combustible liquids in containers ranging from 1 pt aerosol cans up to, and including, 55-gal metal drums. Method of storage was mostly on wooden pallets on open metal racks, double row with 3 and 4 tiers to a total storage height of 15 ft to 17 ft. A considerable portion of the racks were used for storage of flammable and combustible liquids in 5-gal and 55-gal metal containers on wooden pallets, 4 tiers high. Both flammable and nonflammable aerosols in pint cans in cartons were palletized and stored in portions of the racks. Ceiling sprinkler design was wet pipe, extra-hazardous schedule, using 17/32 orifice, 165°F (73.9°C) heads, supplied from a fairly strong city water supply (52 psi static, 38 psi residual, with 1,580 gpm flowing). A review of the hydraulics indicates system was capable of supplying a density of 0.20 gpm/sq ft for the most remote 2,000 sq ft area.

Despite immediate fire department response to a central station water flow alarm and use of a fire department siamese connection, the fire spread beyond the capability of the sprinkler system and the system was soon overtaxed, resulting in early roof collapse and breaking of sprinkler piping, and thus requiring closing of the main control valve. Numerous "fire ball" explosions of aerosol cans and ruptures of 55-gal drums were reported, several affecting manual fire fighting operations, requiring about 5 hr for control.

(5) **1975 Fire.** About one hundred 55-gal drums of Class IB and IC liquids were stored palletized, 3 drums high, in a corner of a general purpose warehouse together with ordinary combustible commodities up to 11 ft high in racks. The roof was Class II steel deck, 15 ft high.

Sprinklers were on an ordinary hazard system, 160°F (71.1°C) heads.

Employees discovered a large fire in progress in the drum storage area. Shortly after the public fire department arrived, drums started to rupture creating large fire balls. One drum failed at the bottom and rocketed through the roof, landing 750 ft from the building. The roof partially collapsed and one system was then shut off. Most of the building and contents were severely damaged.

The fire probably started in an open waste pail near the drum storage. Total loss was about $3,300,000.

SI Units

1 gal = 3.785 L
1 ft   = 0.3048 m
1 sq ft = 0.0929 m^2

Appendix E

Referenced Publications

E-1 This portion of the Appendix lists publications referenced within this NFPA document, and thus is considered part of the requirements of the document.

E-1-1 NFPA Publications. The following publications are available from the National Fire Protection Association, Batterymarch Park, Quincy, MA 02269.

NFPA 11-1978, *Standard for Foam Extinguishing Systems*

NFPA 12-1980, *Standard on Carbon Dioxide Extinguishing Systems*

NFPA 12A-1980, *Standard on Halon 1301 Fire Extinguishing Systems*

NFPA 13-1980, *Standard for the Installation of Sprinkler Systems*

NFPA 15-1979, *Standard for Water Spray Fixed Systems for Fire Protection*

NFPA 17-1980, *Standard for Dry Chemical Extinguishing Systems*

NFPA 45-1975, *Fire Protection for Laboratories Using Chemicals*

NFPA 69-1978, *Standard on Explosion Prevention Systems*

NFPA 70-1981, *National Electrical Code*

NFPA 80-1979, *Standard for Fire Doors and Windows*

NFPA 87-1980, *Standard for the Construction and Protection of Piers and Wharves*

NFPA 90A-1978, *Standard for the Installation of Air Conditioning and Ventilating Systems.*

NFPA 91-1973, *Standard for the Installation of Blower and Exhaust Systems for Dust, Stock, and Vapor Removal or Conveying*

NFPA 101-1981, *Life Safety Code*

NFPA 231-1979, *Standard for Indoor General Storage*

NFPA 231C-1980, *Rack Storage of Materials*

NFPA 251-1979, *Standard Methods of Fire Tests of Building Construction and Materials*

NFPA 302-1980, *Fire Protection Standard for Motor Craft*

NFPA 303-1975, *Fire Protection Standard for Marinas and Boatyards*

NFPA 321-1976, *Basic Classification of Flammable and Combustible Liquids*

NFPA 329-1977, *Underground Leakage of Flammable and Combustible Liquids*

NFPA 385-1979, *Standard for Tank Vehicles for Flammable and Combustible Liquids*

NFPA 386-1979, *Standard for Portable Shipping Tanks for Flammable and Combustible Liquids*

NFPA 496-1974, *Standard for Purged Enclosures for Electrical Equipment in Hazardous Locations*

NFPA 704-1980, *Identification of the Fire Hazards of Materials*

E-1-2 Other Publications

ASTM Publications are available from the American Society for Testing and Materials, 1916 Race Street, Philadelphia, PA 19103

ASTM D86-78, *Standard Method of Test for Distillation of Petroleum Products*

ASTM D56-79, *Standard Method of Test for Flash Point by the Tag Closed-Cup Tester*

ASTM D93-79, *Standard Method of Test for Flash Point by the Pensky-Martens Closed Tester*

ASTM D3243-77, *Standard Methods of Tests for Flash Point of Aviation Turbine Fuels by Setaflash Closed Tester*

ASTM D3278-73, *Standard Method of Tests for Flash Point of Liquids by Setaflash Closed Tester*

ASTM D5-73(1978), *Test for Penetration for Bituminous Materials*

ASTM D323-79, *Standard Method of Test for Vapor Pressure of Petroleum Products (Reid Method)*

ASTM A395-77, *Ferritic Ductile Iron Pressure Retaining Castings for Use at Elevated Temperatures*

ASTM D92-78, *Cleveland Open-Cup Test Method*

ASTM/ANSI D3435-78, *Plastic Containers (Jerry Cans) for Petroleum Products*

ANSI and ASME Publications are available from the American Society of Mechanical Engineers, United Engineering Center, 345 East 47 Street, New York, NY 10017

ANSI B31, *American National Standard Code for Pressure Piping*

ASME, *Boiler and Pressure Vessel Code*

API Publications are available from the American Petroleum Institute, 2101 L Street, N.W., Washington, DC 20037

API 650, *Welded Steel Tanks for Oil Storage*, Sixth Edition, 1978

API Specifications 12B, *Bolted Tanks for Storage of Production Liquids*, Twelfth Edition, January, 1977

API 12D, *Field Welded Tanks for Storage of Production Liquids*, Eighth Edition, January, 1977

API 12F, *Shop Welded Tanks for Storage of Production Liquids*, Seventh Edition, January, 1977

API 620, *Recommended Rules for the Design and Construction of Large, Welded, Low-Pressure Storage Tanks*, Fifth Edition, 1973

API 2000, *Venting Atmospheric and Low Pressure Storage Tanks*, 1973

API 1615, *Installation of Underground Petroleum Storage Systems*, 1979

API 1621, *Recommended Practice for Bulk Liquid Stock at Retail Outlets,* 1977

UL Publications are available from Underwriters Laboratories Incorporated, 333 Pfingsten Road, Northbrook, IL 60062

UL 142-1972, *Standard for Steel Aboveground Tanks for Flammable and Combustible Liquids*

UL 80-1974, *Standard for Steel Inside Tanks for Oil Burner Fuel*

UL 842-1974, *Standard for Valves for Flammable Fluids*

STI P3-1980, *Specifications for STI-P3 System of Corrosion Protection of Underground Steel Storage Tanks*, available from Steel Tank Institute, 666 Dundee Road, Suite 705, Northbrook, IL 60062

E-2 This portion of the Appendix lists publications which are referenced within this NFPA document for information purposes only, and thus is not considered part of the requirements of the document.

E-2-1 NFPA Publications. The following publications are available from the National Fire Protection Association, Batterymarch Park, Quincy, MA 02269.

NFPA 10-1978, *Standard for Portable Fire Extinguishers*

NFPA 14-1980, *Standard for the Installation of Standpipes and Hose Systems*

NFPA 24-1977, *Standard for Outside Protection*

NFPA 31-1978, *Standard for the Installation of Oil Burning Equipment*

NFPA 32-1979, *Standard for Drycleaning Plants*

NFPA 33-1977, *Standard for Spray Application*

NFPA 34-1979, *Standard for Dipping and Coating Processes*

NFPA 35-1976, *Standard for the Manufacture of Organic Coatings*

NFPA 36-1978, *Standard for Solvent Extraction Plants*

NFPA 37-1979, *Standard for Installation and Use of Stationary Combustion Engines and Gas Turbines*

NFPA 51-1977, *Standard for the Installation and Operation of Oxygen-Fuel Gas Systems for Welding and Cutting*

NFPA 56C-1980, *Standard for Laboratories in Health Related Institutions*

NFPA 68-1978, *Guide for Explosion Venting*

NFPA 71-1977, *Standard for the Installation, Maintenance and Use of Central Service Signaling Systems for Guard, Fire Alarm and Supervisory Service*

NFPA 72A-1979, *Standard for the Installation, Maintenance and Use of Local Protective Signaling Systems for Watchman, Fire Alarm, and Supervisory Service*

NFPA 72B-1979, *Standard for the Installation, Maintenance and Use of Auxiliary Protective Signaling Systems for Fire Alarm Service.*

NFPA 72C-1975, *Standard for the Installation, Maintenance and Use of Remote Station Protective Signaling Systems*

NFPA 72D-1979, *Standard for Installation, Maintenance and Use of Proprietary Protection Signaling System for Watchman, Fire Alarm and Supervisory Service*

NFPA 77-1977, *Recommended Practice on Static Electricity*

NFPA 78-1980, *Lightning Protection Code*

NFPA 204-1968, *Guide for Smoke and Heat Venting*

NFPA 327-1975, *Standard Procedures for Cleaning or Safeguarding Small Tanks and Containers*

NFPA 395-1980, *Standard for the Storage of Flammable and Combustible Liquids on Farms and Isolated Contruction Projects*

NFPA 1221-1980, *Public Fire Service Communications*

Index

— A —

Aboveground tanks, spacing between
 aboveground tanks and spillage control..2-2.3
 Class I, II or IIIA liquids ...2-2.2.1, Table 2-7
 Class IIIB liquids....................2-2.2.3
 crude petroleum tanks 2-2.2.2

Aboveground tanks, venting for
 in tanks storing Class I, II and III
  liquids.........................2-2.5.2
 in vertical tanks....................2-2.5.3
  (see Appendix A)
 low pressure tanks, vents for.........2-2.4.3
 pressure vessels, vents for...........2-2.4.3
 venting devices, conditions for
  omissions......................2-2.4.7
 vents, drains and outlets............2-2.4.5
 venting purposes...................2-2.4.1
 vents, restrictions on................2-2.4.6
 vents, sizes and flow................2-2.4.4
 vents, standards for.................2-2.4.2

Aboveground tanks, vent piping for
 additional vents, required for 2-2.7.1
 additional vents, restrictions on 2-2.7.2.3
 fillpipes, construction and placement 2-2.7.4.5
 vent pipe outlets, limits on
  manifolding of 2-2.6.3.4
 vent pipe outlets, location of 2-2.6.2
 exceptions to spacing patterns 2-2.2.5
 impounding around tanks by diking ...2-2.3.3
 liquid storage tanks and LP-Gas
  containers, spacing 2-2.2.6
 minimum tank spacing 2-7
 remote impounding, drainage system ..2-2.3.2
 unstable liquids 2-2.2.4

— B —

Bottom loading tank vehicles
 connections and vapor control 6-3.4.3.5
 vapor processing equipment 6-3.6
 with or without vapor control 6-3.4.1
 with vapor control...................6-3.4.2

Buildings
 exits 6-2.1
 heating, restrictions on...............6-2.2
 loading and unloading facilities,
  location and equipment 6-3.1.2
 ventilation, requirements for.......6-2.3.1.2.3

Bulk plants, electrical equipment in
 area classifications 6-5.3.4
 classified areas, location and sizes ...6-1.1.2.3
 drainage and waste disposal 6-7.1
 NEC Class I........................6-5.3
 sources of ignition, protection against ...6-6.1
 specifications for 6-5.2

Bulk plants, fire control in
 emergency procedure plan, written and
  posted..........................6-8.3
 fire protection facilities 6-8.2
 portable fire extinguishers............6-8.1

Bulk plants and terminals
 piping, valves, and fittings.............6-1.4
  (see Ch. 3)
 storage, requirements for.........6-1.1.2.3.4

— C —

Codes for special installations
 dip tanks 1-1.8
 drycleaning plants....................1-1.8
 fire protection for chemical laboratories .1-1.8
 laboratories in health-related institutions 1-1.8
 manufacture of organic coatings 1-1.8
 solvent extraction plants 1-1.8
 spray applications of flammable and
  combustible materials 1-1.8
 stationary combustion engines and gas
  turbines 1-1.8

Container and portable tank storage
 containers, liquids and flammable
  aerosols.......................4-1.1.2.3
 design, construction, and capacity...4-2.1.2.3
  (see Table 4-2.3)
 exceptions to requirements 4-2.3.1.2

**Container and portable tank storage,
fire control in**
 control of ignition sources in 4-7.2.3.4
 fire extinguishers and hose lines,
  requirements for 4-7.1.1.2.3.4

Cutoff rooms and attached buildings
 basement storage areas 4-4.2.12
 construction standards for 4-4.2.1
 dispensing operations, restrictions on.4-4.2.11
 drainage system for..................4-4.2.5
 fire door in 4-4.2.2
 fire resistance rating for 4-4.2.3
 indoor unprotected storage,
  requirements for 4-4.2.7
 openings.........................4-4.2.4

248   FLAMMABLE AND COMBUSTIBLE LIQUIDS CODE HANDBOOK

protected storage 4-4.2.8
    (see Section 4-6 and Appendix C)
roofs for 4-4.2.6
storage quantities, standards for 4-4.2.10
unprotected storage 4-4.2.7
    (see Table 4-4.2.7)
use of wood in 4-4.2.9

— D —
Definitions
of aerosol 1-2
of apartment house 1-2
of approved 1-2
of assembly occupancy 1-2
of atmospheric tank 1-2
of attached building 1-2
of authority having jurisdiction 1-2
of automotive service stations 1-2
of barrel 1-2
of basement 1-2
of boiing point 1-2
of boil-over 1-2
of bulk plant or terminal 1-2
of chemical plant 1-2
of closed container 1-2
of combustible liquid 1-2
of container 1-2
of crude petroleum 1-2
of cutoff room 1-2
of distillery 1-2
of dwelling 1-2
of dwelling unit 1-2
of educational occupancy 1-2
of fire area 1-2
of flammable aerosol 1-2
of flammable liquid 1-2
of flash point 1-2
of hotel 1-2
of inside room 1-2
of institutional occupancy 1-2
of labeled, authorized equipment 1-2
of liquid 1-2
of listed, authorized equipment 1-2
of low pressure tank 1-2
of marine service station 1-2
of office occupancy 1-2
of portable tank 1-2
of pressure vessel 1-2
of protection for exposures 1-2
of refinery 1-2
of safety can 1-2
of separate inside storage area 1-2
of service station inside buildings .. 1-2
of unstable (reactive) liquid 1-2
of vapor pressure 1-2
of vapor processing equipment 1-2
of vapor processing system 1-2
of vapor recovery system 1-2
of ventilation 1-2

of warehouse, general purpose 1-2
of warehouse, liquid storage 1-2

Drainage systems
emergency drains, restrictions on 7-5.5.2
requirements for 7-5.5.1

— E —
Emergency venting requirements
exceptions to 2-2.5.2
for tanks over 2,800 square feet 2-8
    (see Appendix A)
for tanks under 2,800 square feet 2-8
    (see Appendix A)
in aboveground storage tanks 2-2.5.1
in vertical tanks, types of 2-2.5.3
pressure relieving devices, standards
  for 2-2.5.4
rate of venting, formula for 2-2.5.5
    (see Appendix A)
stable liquid tanks adjustments to
  venting capacity 2-2.5.7
stable liquids, determining relief,
  venting for 2-2.5.6
tank venting devices, marked for
  pressured and flow capacity 2-2.5.9
tank venting devices, marking methods
  for 2-2.5.9
tank venting devices, formula for flow
  tests 2-2.5.9
vents and vent drains, placement of .. 2-2.5.8
wetted area and tank size 2-2.5.4
    (see Appendix A)
**Exceptions to Flammable and
Combustible Liquids Code**
halogenated hydrocarbons and mixtures 1-1.7.4
mists, sprays, or foams 1-1.7.3
storage and use of containers 1-1.7.2
storage on construction projects ... 1-1.7.3
storage on farms 1-1.7.3
transportation of 1-1.7.1
Exits
requirements for 1-5.1

— F —
Flammable and Combustible Liquids Code
alterations in *Code* 1-1.5
application and scope of 1-1.1
electrical classification of 1-1.3
non-compliance to *Code* 1-1.5
storage requirements of 1-1.2
unusual burning characteristics of .. 1-1.4
ventilation of 1-1.3
volatility of 1-1.3

Fuel dispensing units
Class I liquids, dispensing devices .. 7-4.3.1.2

INDEX 249

collision damage, protection against . . . 7-4.3.5
dispensing devices, location of 7-4.1
dispensing liquids, methods allowed . . . 7-4.3.4
emergency power cutoff 7-4.1.2
manual switch control 7-4.3.3

— H —

Housekeeping
aisles, requirements for 5-9.2
classified areas, locations and sizes 5-7.3
combustible wastes and grounds 5-9.3.4
maintenance practices, leakage control . . 5-9.1

— I —

Ignition
sources of 5-6.1
static electricity, prevention of 5-6.2
(See also 1-1.3)

Indoor storage
assembly occupancies and hotels 4-5.3
basement storage areas 4-5.6.4
basic conditions for 4-5.1.1.2.3
dwellings and residential buildings 4-5.2
mercantile occupancies and
  retail stores 4-5.5.1.2.3.4.5.6
mixed storage 4-5.6.7
office, educational, and
  institutional occupancies 4-5.4.1.2.3.4
palletized, solid pile, or rack
  storage 4-5.6.5.6
(see Appendix C)
warehouse, general purpose 4-5.6.1.2.3
(see Appendix D)

Industrial plants
cleaning fluids production,
  requirements for 5-3.1.2.3
cosmetics production, requirements
  for 5-3.1.2.3
drainage, standards for 5-3.4.1.2.3
(see also 2-2.3.1)
handling liquids at point of final
  use in 5-2.4.1.2.3.4.5
incidental storage or use in,
  requirements 5-2.1.2.3
insecticides production, requirements
  for 5-3.1.2.3
pharmaceuticals production,
  requirements for 5-3.1.2.3
solvents production, requirements for 5-3.1.2.3
tank vehicles and tank cars, loading
  and unloading 5-4.1
types of 5-1.1.2
ventilation, equipment and requirements
  for 5-3.1.2
(see also Section 8-4)

Industrial plants, electrical equipment in
in classified areas 5-7.3.4.5.6
(see Table 5-7.3)
in liquid storage areas, restrictions on . . 5-7.1.2

Industrial plants, fire control in
fire alarm system 5-5.5
inspection, requirements 5-5.6
portable fire control equipment 5-5.1
special equipment, requirements for . . . 5-5.3.4
water supply 5-5.2

Inside buildings
design, construction, test and inspection
  of 7-3.1
(see Ch. 3)
liquid storage, restrictions on 7-2.3.1.2.3
piping 7-3.1.2
shutoff valve location of 7-3.1.3
stray currents 7-3.1.1
testing for maximum operating
  pressure 7-3.1.4

Inside installation of tanks
additional vents, required for 2-4.4.1.2
additional vents, restrictions on 2-4.4.2.3
fill pipes, construction and
  placement 2-4.4.4.5.6.7
location of restrictions on 2-4.1
(see also Ch. 5, 6, 7, 8, or 9)
overflow devices, types and
  requirements for 2-4.4.8.9
restrictions on vents for 2-4.2.3

Inside rooms
aisle, requirements for 4-4.1.7
basement storage 4-4.1.9
construction standards for 4-4.1.1
dispensing procedure 4-4.1.8
(see also Ch. 5)
electrical wiring and equipment for 4-4.1.5
openings, standards for 4-4.1.2
storage, requirements for 4-4.1.4
ventilation for 4-4.1.6

— L —

Liquid storage, protection for
protected, palletized, or solid
  pile storage . . . . Table 4-6.1(a)
protected rack storage . . . . Table 4-6.1(b)
requirements 4-6.1.1.2.2.1.3.4.5
(see Appendix C)

Liquid warehouses
attached warehouses, standards for . . . 4-5.7.4
basement storage 4-5.7.7
container placement, location
  and spacing within 4-5.7.8.9.10.11.12

location and spacing............4-5.7.1.2.3
mixed storage....................4-5.7.13
  [see Table 4-6.1(b)]
storage quantities, restrictions for.....4-5.7.6

Loading and unloading tank vehicles,
  requirements for...................8-5.1
  (see Section 6-3)

— O —

Outdoor liquid storage
  location and maximum amounts,
    guide to....................4-8.2.1.2.3.4
  requirements for 4-8.1.1.2.3.4

Outside aboveground tanks
  characteristics of 2-2.1.1
  location and important buildings 2-2.1.1
    (see Tables 2-1 through 2-6)
  location and property lines...........2-2.1.1
    (see Tables 2-1 through 2-6)
  location and public ways 2-2.1.1
    (see Tables 2-1 through 2-6)
  location of 2-2.1.1
  property line rules, exceptions to ..2-2.1.7.1.8
  restrictions on aboveground tanks .2-2.1.3.1.6
  tanks for liquids with boil-over
    characteristics 2-2.1.4
  tanks for liquids with unstable liquids .2-2.1.5
  vertical tanks, restrictions on 2-2.1.2

— P —

Pipe joints
  corrosion, protection against 3-5.1
  requirements for 3-3.1.2
  supports, protection against stress and
    damage...........................3-4.1
  testing, standards for 3-7.1
  valves, requirements for control of
    flow 3-6.1

Piping
  requirements for 7-5.4.1
    (see Ch. 3)
  vertical shafts and horizontal chases ...7-5.4.2

**Piping, valves and fittings, design and
  construction**
  low melting point materials, guide to ..3-2.4.5
  major parts of 3-1.3
  materials for tanks, guide to 3-2.3.2
  nodular iron, standards for 3-2.2
  other materials for valves, guide to 3-2.3.1
  piping systems, materials for..........3-2.1
  requirements for.....................3-1.1
  steel or iron valves, standards for 3-2.3

Piping, valves, and fittings, requirements
  listed flexible connectors...........8-4.2.1.2
  marked piping 8-4.2.3

transfer of liquids, approved methods
  for 8-4.3.1.2
vapor control, protection for 8-4.4.1.2

Portable containers
  dispensing, restrictions on 7-8.2
  sales, restrictions on 7-8.2.1

Processing plants
  construction standards for 8-3.1.1
  emergency drainage systems,
    requirements for..............8-3.2.1.2.3
  exit facilities.......................8-3.1.2
  explosion relief, guide for............8-3.4.1
  processing vessels, location and spacing 8-2.1.1
    (see Table 8-2.1)
  special chemical plants, types of 8-1.1
  ventilation, natural or mechanical,
    requirements for.................8-3.3.1.2

**Processing plants, electrical
  equipment in**
  classified areas, definitions of ...8-7.3.3.4.5.6
    (see Table 8-7.3)
  classified areas, guide to 8-7.3
  requirements for..................8-7.3.1.2

Processing plants, fire control in
  automatic sprinkler system...........8-6.2.4
    (see Appendix C)
  fire alarm system....................8-6.3
  fire protection facilities, maintenance and
    inspection of 8-6.4
  hose, installation of.................8-6.2.3
  portable fire extinguishers 8-6.1
  special hazards 8-6.2
  water supply and hydrants 8-6.2.2.3

Processing plants, housekeeping in
  aisles...............................8-8.2
  maintenance and spillage control 8-8.1
  waste disposal and grounds
    maintenance......................8-8.3.4

Processing plants, liquid handling in
  inside storage tanks, requirements for 8-4.1.3.4
    (see Ch. 4)
  storage, requirements for 8-4.1.12
    (see Ch. 2)

— R —

Refineries, chemical plants and distilleries
  location 9-1.2
    (see also 2-2.1.2)
  piping, valves and fittings 9-1.3
    (see also Ch.3)
  process units, location of 9-3.1

INDEX 251

storage, types of 9-1.1
    (see also Ch. 2)
wharves, requirements for 9-2.1
    (see also Section 6-4)

Refineries, chemical plants, distilleries, fire control in
emergency control group, requirements
    for............................... 9-4.8
fire alarm system, requirements for ... 9-4.7.9
hot work, restrictions on and inspection
    of 9-4.2
maintenance, leakage control 9-4.3
portable fire extinguisher equipment 9-4.4
smoking, restrictions on............... 9-4.1
special extinguishing equipment 9-4.6
    (see Appendix C)
water supply 9-4.5

Remote pumping systems
Class I liquids..................... 7-3.2.1
design and equipment............... 7-3.2.2
location of 7-3.2.3
shutoff valves, requirements of 7-3.2.5
submersible pumps, piping manifolds . 7-3.2.4
subsurface pumps, pits for........... 7-3.2.4
vapor return pipe, requirements for ... 7-3.2.6

Repairs to equipment
hot work operations, inspection
    requirements 5-8.1

— S —

Self-service stations
attendant, requirements for....... 7-8.4.3.4.7
drainage and waste control 7-8.5.1.2
emergency controls 7-8.4.5
fire control, requirements for 7-10.1
operating instructions for, posting
    required 7-8.4.6
requirements for.................. 7-8.4.1.2
sources of ignition, restrictions for...... 7-9.1
warning signs for, posting required.... 7-8.4.8

Service stations
aboveground tanks, dispensing
    methods 7-2.1.2
automotive, marine, and indoor........ 7-1.1
Class I liquids and ventilated pits 7-2.1.3
indoor service stations, tank limits 7-2.2.3
installation of tanks, in buildings.... 7-2-2.1.2
leakage control, inventory records as
    aids to 7-2.1.4
piping and hose failure, protection for . 7-2.1.6
    (see also 2-2.7.1)
shore tanks, location of 7-2.1.5
special enclosures 7-2.1.1
    (see 7-8.3.4 and 7-8.3.5)
storage containers, types of 7-2.1.1
    (see 7-8.3.4 and 7-8.3.5)
    (see Appendix D)

Service stations, electrical equipment in
classified areas, requirements for 7-6.3.4
    (see Table 7-5)
liquid storage, requirements for 7-6.1.2

Service stations inside buildings
construction and location of 7-5.1.2.3.4
dispensing area at street level 7-5.2.1
fire doors in 7-5.1.5
number of vehicles, serviced by 7-5.2.2
openings, protection of............. 7-5.1.6

Service stations, operational requirements for
classified areas, electrical equipment for
    service stations 7-1
Class I, Class II, Class III liquids,
    marked containers for............. 7-8.2.1
fuel delivery nozzles, requirements
    for Class I liquids 7-8.1
hose nozzle valves 7-8.1.2.3.4.5

Service stations, supervision of dispensing
Class I and Class II liquids
    dispensing restrictions............. 7-8.3.5
private locations, exceptions for 7-8.3.3.4
self-service stations, restrictions on.... 7-8.3.2
service stations, supervisors for....... 7-8.3.1

Sources of ignition
maintenance and repair............ 8-7.2.1.2
precautions for................... 8-7.1.1.2
    (see 1-1.3)

Static protection
bonding facilities, exceptions to 6-3.7.3
bonding facilities, requirements for.. 6-3.7.1.2
container filling facilities, bonding
    requirements for 6-3.9
downspouts 6-3.7.4
stray currents...................... 6-3.8

Storage cabinets
design, construction and capacity 4-3.1.2
metal cabinets 4-3.2.1
wooden cabinets 4-3.2.2
listed cabinets..................... 4-3.2.3

— T —

Tanks, requirements for all types
aboveground tanks, special
    restrictions................... 2-5.6.1.2.3.4
anchorage for 2-5.6
emergencies, instructions and
    procedures for 2-5.6.4.1.2-5.7
ignition, sources of.................. 2-6.1
fire protection and identification...... 2-8.1.2
    (see Appendix A)
foundations for 2-5.1.2

overfilling of tanks, prevention and
  detection of 2-9.1.10.1
supports for 2-5.3.4.5
testing, requirements for 2-7.1.2.3.4.5
underground tanks, special
  restrictions 2-5.6.2.1.2.3

Tank storage, design and construction
atmospheric tanks 2-1.3.1.3.2.3.3
construction methods 2-1.2.2
fabrication of tanks................. 2-1.2.1
low pressure tanks 2-1.4.1.4.2.4.3
materials, choice of in tank
  construction 2-1.1
metal tanks....................... 2-1.2.2
provisions for internal corrosion 2-1.6.1
restrictions on pressure vessels 2-1.5.1.5.2
restrictions on tanks for combustible
  liquids 2-1.1
      (see Appendix D)
shape of tanks 2-1.2.1

Tank storage, spacing
boil-over liquids .... 2-2.1.4(a)
      (see Table 2-3)
Class III B liquids .... Table 2-5
stable liquids at greater than 2.5 psig .. 2-2.2-6
      (see Table 2-2)
stable liquids at 2.5 psig or less
  .... 2-2.1.1
      (see Table 2-1)
unstable liquids 2-2.2.4
      (see Table 2-4)

Top loading tank vehicles
with vapor control................. 6-3.3.2
without vapor control.............. 6-3.3.1

— U —

Underground tanks
burial depth and cover for.......... 2-3.2.1.2
disposal, abandonment, or removal of .2-3.4.1
      (see Appendix B)
installation of 2-3.2.3.4
location of 2-3.1

Underground tanks, vent piping for
additional vents, restrictions on 2-3.6.3.4
additional vents, required for 2-3.6.1.2
Class II or Class IIIA liquids, location
  of vents for..................... 2-3.5.3
fill pipes, construction and location .. 2-3.6.5.6
vents, capacity of.................. 2-3.5.2
      (see Table 2-10)
vent line diameters.............. Table 2-10
vents, location and arrangement of 2-3.5.1
      (see also 7-2.1.1)

vent piping, construction of 2-3.5.4
      (see also Ch. 3)
vent pipe outlets, limits on
  manifolding of 2-3.5.5.6

— V —

Vapor processing systems
blower-assist, restricted use of 7-4.5.4
combustion or open flame devices.... 7-4.5.10
      (see Table 7-5)
components, listing of 7-4.5.1
dispensing devices, listing of 7-4.5.2
electrical equipment 7-4.5.8
      (see Table 7-5)
internal explosions, prevention of 7-4.5.5
vapor discharge, prevention of 7-4.5.3
vapor processing equipment, location
  of 7-4.5.6.7
vents and outlets, placement of 7-4.5.9

Vapor recovery systems
hose nozzle valves 7-4.4.3
listed dispensing devices, modification
  requirements for 7-4.4.2
vapor control 7-4.4.4

Ventilation and heating equipment
exceptions to requirements........... 7-5.3.6
installation, requirements for ... 7-1.2.3.4.5.6
mechanical exhaust system,
  requirements for 7-5.3.2.3.4.5
separate heating, ventilating, and air
  conditioning systems.............. 7-5.3.1

— W —

Wharves
cargo transfer and other mechanical
  work 6-4.7.1
cargo transfer, supervision of 6-4.7
Class I and Class II liquids, special
  requirements for 6-4.5.4
definition and liquid cargo transfers ... 6-4.1.1
design and construction of............ 6-4.2
location and other structures 6-4.1.2
fire extinguishers and fire hoses....... 6-4.6.1
hoses, mooring lines and vessel
  surge 6-4.5.7.8
housekeeping and access roads...... 6-4.6.2.3
loading pumps on 6-4.4
pipe lines, bonding and grounding of .. 6-4.5.6
piping, valves, and fittings on 6-4.5.1.2.3
      (see Ch. 3)
pressure hoses and couplings, use on .. 6-4.4.1
tank storage on 6-4.3